OYSTER PERPETUAL

SUBMARINER

The Watch that Unlocked the Deep

ROLEX

OYSTER PERPETUAL

SUBMARINER

The Watch that Unlocked the Deep

NICHOLAS FOULKES

Wallpaper*

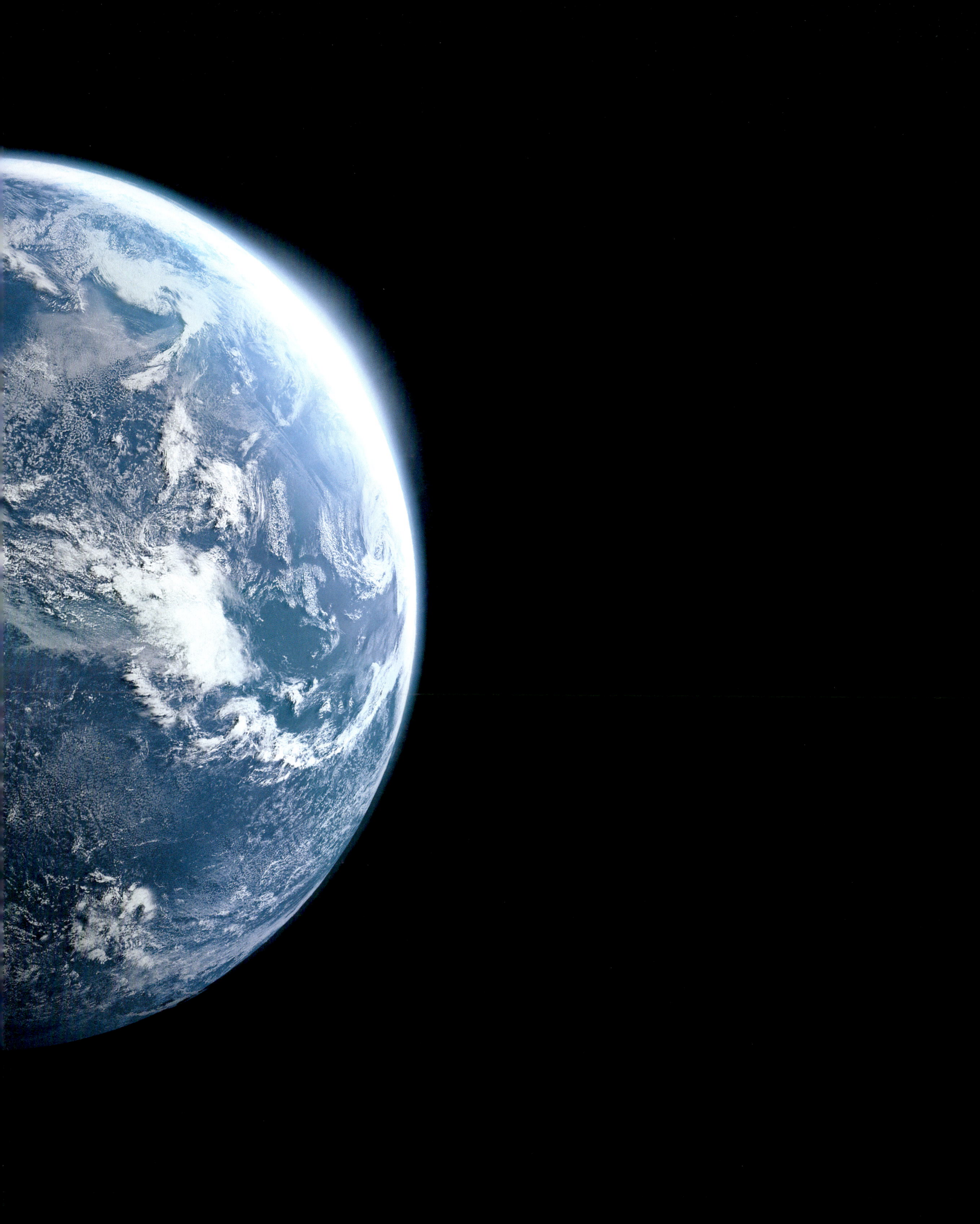

50
40

ROLEX
SUBMARINER
1000ft = 300m
SUPERLATIVE CHRONOMETER
OFFICIALLY CERTIFIED

Previous pages: Our unique world is cloaked by the oceans, the cradle of all life and the natural environment of the Submariner. Pictured are: Three Rooker Island on Florida's Gulf Coast, now part of a Hope Spot created by Mission Blue, a partner of Rolex's Perpetual Planet Initiative; the latest model of the Submariner, launched in 2020; and marine conservationist and Rolex Awards for Enterprise Laureate Brad Norman with his Submariner in Western Australia.

CONTENTS

“ROLEX MUST CONTINUOUSLY STRIVE TO THINK AND ACT DIFFERENTLY FROM THE REST. THEREIN LIES MY GREATEST STRENGTH.”

HANS WILSDORF, *Founder of Rolex*

Hans Wilsdorf, 1905.

THE VISION OF HANS WILSDORF

One morning in 1908, "while riding on the upper deck of a horse-drawn omnibus along Cheapside in the City of London, a genie whispered 'Rolex' in my ear."[1]

The ear into which that supernatural being whispered belonged to 27-year-old Hans Wilsdorf, one of the most remarkable entrepreneurs of the 20th century. His influence on the field of watchmaking can be likened to that of Henry Ford on the motor car. A pioneer in moving the watch from the pocket to the wrist, Wilsdorf created one of the defining objects of the modern world and the strange-sounding name that occurred to him on that bus journey is now, well over 100 years later, known in almost every corner of the world.

Wilsdorf devoted his life to the mission of proving that the wristwatch could be as precise and dependable as the pocket watch. His first great breakthrough came on the eve of World War One when, in July 1914, "the first Class 'A' Observatory Certificate was awarded to a wrist-watch, namely a 25 mm Rolex movement. The success of this test gave a great impetus to the development of the wrist-watch."[2]

The next momentous step in the journey was taken in 1926 with the arrival of the Rolex Oyster, so named because it was totally hermetic, sealing the precious movement off from the deleterious effects of dust and water. The Oyster's remarkable properties were promoted by cross-Channel swimmer Mercedes Gleitze in 1927.

A further triumph came in 1931 with the launch of the Rolex Oyster Perpetual, which used the wearer's movements to wind the watch, removing the risk of water or dust entering the timepiece during winding.

In the 1950s, Rolex launched its "Professional" watches. These rugged, functional products include some of the most famous watches ever made: the Explorer, the Submariner, the GMT-Master and the Milgauss.

In the second half of the century, the name created by that Edwardian genie became known everywhere. Rolex watches were worn by deep-sea divers and astronauts alike, as well as political leaders as diverse as Sir Winston Churchill and Martin Luther King Jr.

In popular culture, Rolex became indissolubly associated with cinema's Cold War secret agent, James Bond. Sportsmen of the stature of Arnold Palmer, Sir Jackie Stewart and Jean-Claude Killy became Rolex Testimonees. Even those who would never own a Rolex were to become aware of the name and the crown emblem as the brand sponsored a growing number of world-famous sporting events, including Wimbledon, as well as cultural institutions, such as Teatro alla Scala. Furthermore, the Rolex Awards for Enterprise have supported more than 150 educational, environmental, cultural and scientific endeavours since their creation in 1976.

The 21st century has seen Rolex consolidate its reputation for engineering and industrial excellence with the launch of such models as the highly technical Sky-Dweller and the remarkably robust Rolex Deepsea. The brand has continued to announce ever more prestigious and significant partnerships with, among others, Formula 1® and the Academy of Motion Picture Arts and Sciences. Moreover, the brand's ongoing commitment to understanding and protecting the environment has been given new impetus with the Perpetual Planet Initiative.

The Oyster, the first waterproof wristwatch in the world, was launched in 1926. Its case was hermetically sealed thanks to an ingenious patented system consisting of a screw-down bezel, case back and winding crown.

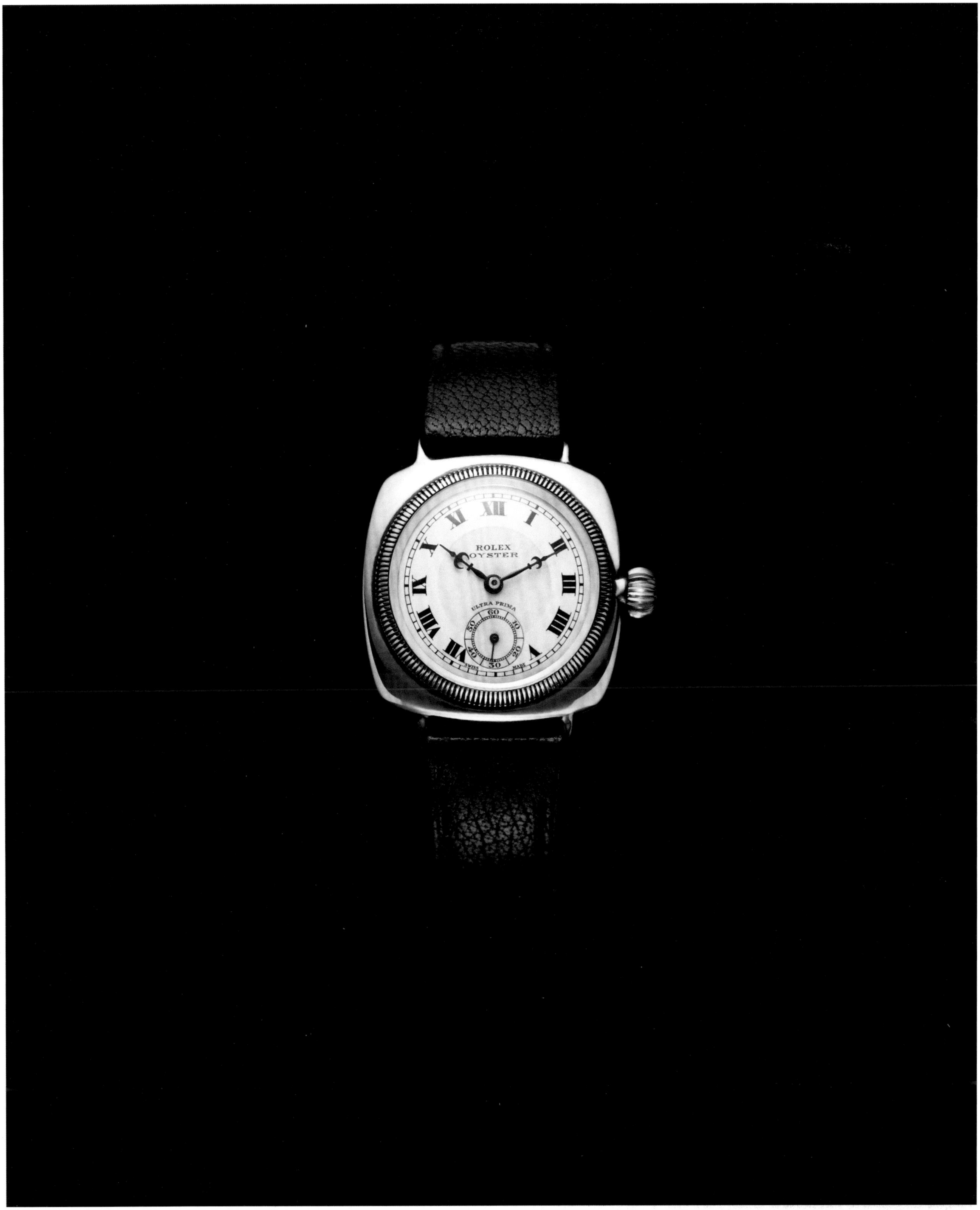
ROLEX
OYSTER
ULTRA PRIMA

Some of the most remarkable figures in history, including Mercedes Gleitze, Martin Luther King Jr and Pablo Picasso, wore Rolex watches.

Equipped for submarine salvage and deepwater training, HMS *Reclaim* entered service with the British Royal Navy in 1948.

A NEW CHAPTER IN WATCHMAKING HISTORY

Loch Fyne is the name given to the 65-kilometre-long gash that bites into the west coast of Scotland. Today, this salt water loch is known for its picturesque vistas and its excellent seafood. But in the 1940s, its now tranquil shores echoed with the rasp and clatter of military convoys, the roar overhead of warplanes, the bark of shouted orders, the detonation of explosions, and the crackle of small arms fire as this remote part of Scotland witnessed some of the most crucial preparations of World War Two.

From 1940 to 1944, a quarter of a million Allied soldiers trained here in the use of landing craft, the tactics of amphibious assault, and the establishment and defence of beachheads. Exercises continued, year-round, during the harsh cold winters as well as the long days of summer, and the lessons learned on Loch Fyne were put into practice in almost every theatre of war, culminating, of course, in the successful D-Day landings of 1944.

By 1946, the naval base was being decommissioned and the many beaches on the shores of the loch returned to their pre-war placidity. Yet, although the need for the mass training of troops for the seaborne assault on Nazi Europe had passed, Loch Fyne still had its military uses as, far from prying eyes, the latest advanced modern warfare techniques were tested in its cold depths.

March is when the waters are at their coldest, seldom topping eight degrees Celsius, and in March 1952, HMS *Reclaim*, a Royal Navy diving support vessel, was lying off the west coast of Scotland conducting training dives in Loch Fyne, where the extreme depths and harsh conditions were considered perfect for the navy's deepwater training. These icy, ink-dark waters were the ultimate proving ground for the navy's elite divers, and the *Reclaim* had been in these waters before: in 1948, the year she entered service, divers based on the *Reclaim* descended to 163 metres (535 feet) below the surface.

Equipped for submarine salvage and fitted with the very latest equipment including underwater television, the *Reclaim* achieved some measure of fame in 1951 for locating the HMS *Affray*, a submarine that had been lost with all hands in the English Channel during a military exercise; it was the first such wreck to be located using television camera technology. Now, on 11 March, 1952, the *Reclaim* received further cutting-edge underwater equipment: "three steel Rolex Oyster Perpetual watches", hand delivered to the remote Scottish loch by Rolex's UK office.

The first mention of this extraordinary military collaboration in the Rolex Archives comes, almost *en passant*, in the PS of a letter written by J. Walter Thompson to Rolex

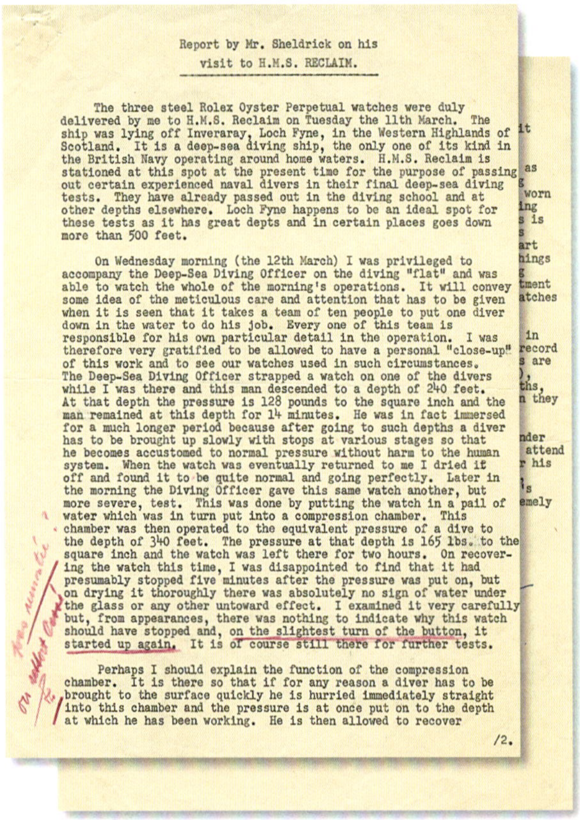

Report by Mr. Sheldrick on his visit to H.M.S. RECLAIM.

The three steel Rolex Oyster Perpetual watches were duly delivered by me to H.M.S. Reclaim on Tuesday the 11th March. The ship was lying off Inveraray, Loch Fyne, in the Western Highlands of Scotland. It is a deep-sea diving ship, the only one of its kind in the British Navy operating around home waters. H.M.S. Reclaim is stationed at this spot at the present time for the purpose of passing out certain experienced naval divers in their final deep-sea diving tests. They have already passed out in the diving school and at other depths elsewhere. Loch Fyne happens to be an ideal spot for these tests as it has great depts and in certain places goes down more than 500 feet.

On Wednesday morning (the 12th March) I was privileged to accompany the Deep-Sea Diving Officer on the diving "flat" and was able to watch the whole of the morning's operations. It will convey some idea of the meticulous care and attention that has to be given when it is seen that it takes a team of ten people to put one diver down in the water to do his job. Every one of this team is responsible for his own particular detail in the operation. I was therefore very gratified to be allowed to have a personal "close-up" of this work and to see our watches used in such circumstances. The Deep-Sea Diving Officer strapped a watch on one of the divers while I was there and this man descended to a depth of 240 feet. At that depth the pressure is 128 pounds to the square inch and the man remained at this depth for 14 minutes. He was in fact immersed for a much longer period because after going to such depths a diver has to be brought up slowly with stops at various stages so that he becomes accustomed to normal pressure without harm to the human system. When the watch was eventually returned to me I dried it off and found it to be quite normal and going perfectly. Later in the morning the Diving Officer gave this same watch another, but more severe, test. This was done by putting the watch in a pail of water which was in turn put into a compression chamber. This chamber was then operated to the equivalent pressure of a dive to the depth of 340 feet. The pressure at that depth is 165 lbs. to the square inch and the watch was left there for two hours. On recovering the watch this time, I was disappointed to find that it had presumably stopped five minutes after the pressure was put on, but on drying it thoroughly there was absolutely no sign of water under the glass or any other untoward effect. I examined it very carefully but, from appearances, there was nothing to indicate why this watch should have stopped and, on the slightest turn of the button, it started up again. It is of course still there for further tests.

Perhaps I should explain the function of the compression chamber. It is there so that if for any reason a diver has to be brought to the surface quickly he is hurried immediately straight into this chamber and the pressure is at once put on to the depth at which he has been working. He is then allowed to recover

/2.

The report by Mr Sheldrick on his visit to HMS *Reclaim*, on which the Royal Navy tested three early models in 1952.

on 29 January 1952. "I am also sending you a suggested letter, drafted by our Public Relations department for you to send to the Commanding Officer of HMS *Reclaim*, the Royal Naval Diving School."[1]

The letter's addressee was one of the most influential executives to work in the Swiss watch industry during the mid-20th century. In 1945, Wilsdorf had made 39-year-old René-Paul Jeanneret a director of Rolex. A former employee of J. Walter Thompson, Jeanneret had a genius for marketing in the broadest possible understanding of the term. He developed advertising campaigns for Rolex that are still remembered today, including, most famously, the strapline "Men Who Guide The Destinies Of The World Wear Rolex Watches." A biographical summary of his work, written for internal purposes in 1973, survives in the company's archives. Far from being an arid document, it is quite unusually effusive in its praise, ending "the Rolex company and perhaps even international advertising would not be what they are now without his 28 years of ceaseless activity."[2]

But more than merely promoting Rolex and its products, he played a crucial role in developing some of the brand's most famous timepieces, chief among them the Submariner, not that it was yet known by that name.

Within a few days, Jeanneret had written to the commanding officer of the Royal Naval Diving School asking for assistance in research "to improve the water-resistant qualities of our wrist-chronometers.

"In our laboratories we endeavour to stimulate [sic], so far as we can, the conditions experienced on the seabed. We find, however, that we obtain our most valuable data from men actually using the Rolex Oyster in their work.

"Our Company feels that occasional reports by Admiralty divers would be of the utmost assistance to us and we are, therefore, hoping that you may allow us to send you two or three of our watches which, with your permission, could be worn – and reported on – by your divers."[3]

Thereafter events moved rapidly, and a little over a month later, Mr Sheldrick of Rolex made the trip with three watches for the divers of HMS *Reclaim*.

In terms of tests carried out under military supervision, the combination of the Royal Navy's crack divers, its most advanced diving support vessel, and the deep, secluded seawater loch was unbeatable. For Rolex, this was a quite remarkable opportunity to subject its watches to in vivo tests at considerable depths on the wrists of the best divers in the world.

For its part, the dive team aboard HMS *Reclaim* was eager to try the new equipment and put the watches to work the following day. Thus, it was on a brisk March morning in an icy Scottish loch, just a few years after the end of World War Two, that the first words were written in a new chapter in watchmaking history as the Royal Navy took its first dive with the watch that would, in time, become the Submariner.

Happily, the event was recorded in detail by Rolex's Mr Sheldrick who, having delivered the watches, had been invited back aboard the *Reclaim* to see the inaugural dive. He described himself as "very gratified to be allowed to have a personal 'close-up' of this work and to see our watches used in such circumstances. The Deep-Sea Diving Officer strapped a watch on one of the divers while I was there and this man descended to a depth of 240 feet. At that depth, the pressure is 120 pounds per square inch (psi) and the man remained at this depth for 14 minutes. He was, in fact, immersed for a much longer period because after going to such depths, a diver has to be brought up slowly with stops at various stages so that his body could get back to normal (or atmospheric) pressure without causing a decompression accident and its potential traumatic consequences. When the watch was eventually returned to me, I dried it off and found it to be quite normal and going perfectly. Later in the morning, the Diving Officer gave this same watch another, but more severe, test. This was done by putting the watch in a pail of water which was in turn put into a compression chamber. This chamber was then operated to the equivalent pressure of a dive to the depth of 340 feet. The pressure at that depth is 165 psi, and the watch was left there for two hours. On recovering the watch this time, I was disappointed to find that it had presumably stopped five minutes after the pressure was put on, but upon drying it thoroughly there was absolutely no sign of water under the glass or any other untoward effect. I examined it very carefully but, from appearances, there was nothing to indicate why this watch should have stopped and, on the slightest turn of the button, it started up again,"[4] suggesting that the watch had simply not been rewound that morning.

From that very first official Royal Navy dive, Rolex started to harvest data about how such a watch would be used and the perils it faced at what were the very limits of human endurance. In the course of preparing the watches for use by the Royal Navy's divers, Rolex had fitted them with extra-long straps so that they could be worn over the cuff of the diving suit. However, Rolex's observer was surprised to learn instead that the watches "will have to be worn on the diver's wrist in the normal way. Specially fitted diving suits allow the divers' hands to be free. The reason for this is because, when they are working very deep, in most waters it is completely dark, and their hands are virtually their eyes. Apart from this, the Diving Officer explained that there are many things such as ropes, etc., which could knock the watch off the diving suit if it was strapped on the outside of the cuff. An assortment of spare straps has now been provided for them to use these watches as they like and to change over as required."[5]

"MEN WHO GUIDE THE DESTINIES OF THE WORLD WEAR ROLEX WATCHES."

RENÉ-PAUL JEANNERET, *Rolex Communications Director*

It was agreed that the ship's officers would keep a detailed log for each watch recording, inter alia, precision, dive depths, and observations about the visibility of the luminous material. The only disappointment was that the *Reclaim's* commanding officer, Lieutenant Commander Bathurst, was not in Scotland but instead was at Buckingham Palace in London to receive an OBE for services relating to the search for the *Affray*.

The tests described by Lieutenant Commander Bathurst in his letter to Rolex of 21 May, 1952 were impressive and detailed, including "having them dived daily over a period of some four weeks to very considerable depths – keeping a daily rate record – and as they stopped when subjected to a certain pressure, actually measuring the compression of the cases by micrometre at varying pressures."[6] However, confusion over whether Rolex had sought the appropriate permission resulted in the research being terminated by the Admiralty, leaving Bathurst "in the position of not being able to furnish you officially with the results of our trials. This is most unfortunate as I collected a considerable amount of data, and was most impressed with the behaviour of the watches under very adverse conditions."[7]

ROYAL NAVAL DIVERS HAVE BEEN ISSUED OFFICIALLY WITH A NEW TYPE OF WRISTWATCH DEVELOPED BY HANS WILSDORF ...

Thus, information that would have been very useful to the development of the Submariner risked becoming, in effect, a military secret. As Jeanneret put it phlegmatically to Richard Winter, the Head of Rolex in London, "it is most regrettable that he should not be unable to give us the benefit of his observations."[8] It took a few frantic weeks of letters and telephone calls before Winter was able to smooth matters over with Bathurst, permitting him to write admonitorily to Jeanneret "with effect from now, if any similar public relations jobs are mooted, and if they are to do with Great Britain or the Empire" (this was 1952 and Britain still had an empire to speak of), "we here in London should be contacted and advised as to what is suggested before any further action is taken."[9] Imperial honour thus restored and the relationship with the Royal Navy back on track, Winter was happy to report that Bathurst "was extremely pleased with the performance of the watches. They have all been down to 350 feet but they all used to stop at depths below 200 feet."[10]

This was no mere PR exercise; it was in fact providing real-world testing data crucial to the development of what would become the most famous diving watch in the world. It was also looking increasingly promising from a commercial perspective as Winter cleverly turned a grovelling letter of apology to the Admiralty into a sales pitch.

"We can fully understand it being contrary to the policy of The Lords Commissioners of the Admiralty to afford special facilities to any one firm unless Naval interests so demand, but we believe that our Company should be included in such a category. We have the honour to suggest that scientifically, hermetically sealed watches could be of great advantage to the Royal Navy whether for use on open ships' decks or below the surface of the sea down to considerable depths and particularly when at such depths, as on the surface, almost chronometric accuracy can be maintained and guaranteed.

"The idea of requesting permission to send the three watches for deep-sea diving tests on a Royal Naval vessel was not so much to assist in our own research work (where, of course, we have machines capable of subjecting our watches to extreme pressures) as to ensure the realization of the hope that our special waterproof Oyster watches might be used by the Royal Navy under extreme conditions with complete confidence.

The British Royal Navy's deep-sea-diving and submarine rescue vessel, HMS *Reclaim* (*bottom*) and HMS *Reclaim* Diving Team (*top*).

As the first divers' wristwatch waterproof to a depth of 100 metres (330 feet), the Submariner, Ref. 6204, 38 mm, marked a major step forward in the history of Rolex and of deep-sea diving.

Opposite: The Royal Navy approved Rolex's public statement deeming the watch waterproof down to 122 metres (400 feet), with dials illuminated for clear reading underwater, 1955.

ROYAL NAVAL DIVERS NOW HAVE THE TIME UNDERWATER

Royal Naval divers have been issued officially with a new type of wrist-watch, developed by Hans Wilsdorf, 74-year old inventor of the Oyster waterproof wrist-watch and the Perpetual self-winding wrist-watch, and Chairman of The Rolex Watch Company Limited.

The watch is self-wound by a rotor mechanism as the diver walks or swims. The dial is a jet black instrument type illuminated for extra clear reading underwater and at night. This official Royal Naval divers' watch was put into production only after two years of exhaustive Admiralty exposure and timing tests in England and the Mediterranean. It has proved itself waterproof even down to 400 feet below the surface of the water. Designated type 6150, the watch is known to the public as "Explorer".

NO ADMIRALTY OBJECTION

CAPTAIN, R.N.

The Rolex Watch Co. Ltd., 1, Green Street, Mayfair, London, W.1.

MAYfair 5071.

23/8/55. JS.

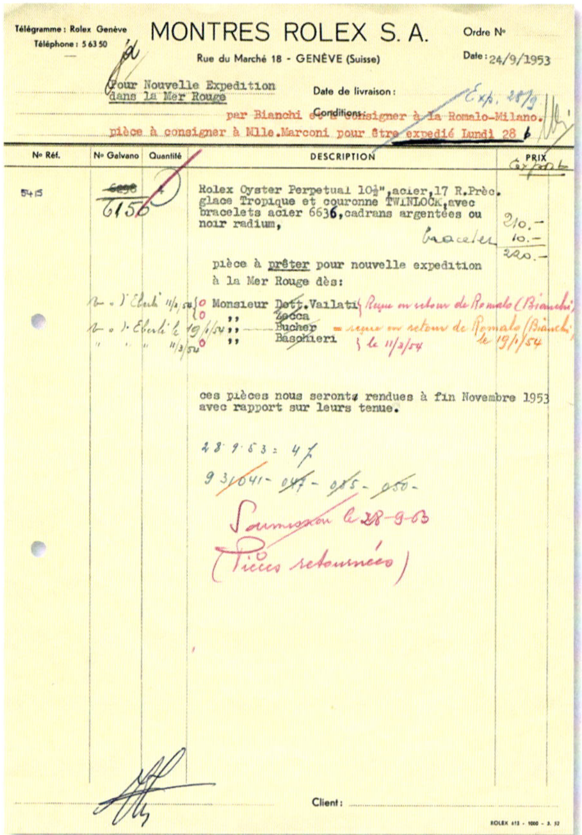

Télégramme : Rolex Genève
Téléphone : 5 63 50

MONTRES ROLEX S. A.

Rue du Marché 18 - GENÈVE (Suisse)

Ordre N°

Date : 24/9/1953

Pour Nouvelle Expedition dans la Mer Rouge

Date de livraison :

Conditions :

par Bianchi et à consigner à la Romalo-Milano.

pièce à consigner à Mlle.Marconi pour être expedié Lundi 28

N° Réf.	N° Galvano	Quantité	DESCRIPTION	PRIX
5415	6156	4	Rolex Oyster Perpetual 10½",acier,17 R.Prèc. glace Tropique et couronne TWINLOCK,avec bracelets acier 6636,cadrans argentées ou noir radium,	210.–
			bracelet	10.–
				220.–
			pièce à prêter pour nouvelle expedition à la Mer Rouge dès:	
			Monsieur Dott.Vailati	
			" Zecca	
			" Bucher	
			" Baschieri	
			ces pièces nous seront rendues à fin Novembre 1953 avec rapport sur leurs tenue.	
			28.9.53 : 47	
			Soumission le 28-9-53 (Pièces retournées)	

Client :

A watch shipping form, September 1953.

Opposite: An early version of the Explorer, Ref. 6150, launched in 1953. Notes in the Rolex Archives detail that it is linked to the Royal Navy.

"We would indeed be deeply appreciative if The Lords Commissioners of the Admiralty would therefore reconsider the position and perhaps, in any case, let us have the full reports (if any) on the watches submitted. We do not know if our watches have been tested but we hereby guarantee and promise that if they have been tested and if reports have been drafted on their performance, we will not use such reports outside the Company's Offices and even inside the Company's Offices, those reports will only be used to assist in correlating further data which may help us in making even further improvements."[11]

The scheme had been conceived as a PR stunt and Winter knew perfectly well that full reports had already been made by Bathurst and remained "in his possession".[12] Moreover, while Rolex doubtlessly possessed effective equipment, the in vivo testing and feedback from highly trained specialist divers was uniquely valuable.

Although Winter's letter might seem disingenuous, it had the desired effect. By September, discussions concerning the Royal Navy's specifications for a watch were advancing apace: HMS *Reclaim* was sending reports on a "new steel Oyster Perpetual", while Rolex London was pressing Geneva for an answer "stating whether you will be able to make the special dial they would require, whether you will be able to include the special anti-magnetic escapement and roughly at the price they would like to pay."[13]

At the same time, Lieutenant Commander Gutteridge of the Underwater Countermeasures and Weapons Establishment was busy conducting anti-magnetic tests on a metal sample supplied by Rolex Bienne. By the autumn of 1952, Gutteridge, an experienced diver and bomb disposal expert in his own right, is described as "anxiously awaiting details concerning the Rolex Oyster Perpetual special watch which may now be considered for purchase by contract by the Admiralty."[14]

On 15 October of that year, Winter was able to write to Gutteridge informing him that watches would be available for testing by the end of November but that in the meantime, "A similar watch equipped with the special dial," but, alas, "without yet the fully non-magnetic escapement, will be in this country towards the end of next week."[15]

Once it had arrived, Winter delivered the watch personally to the Underwater Countermeasures and Weapons Establishment in Hampshire, explaining that it was not the final model. "It has been decided that the radium pointer on the sweep seconds hand should be in the form of a circle enclosing a larger proportion of radium paint, permitting clearer reading of the seconds hand in darkness or semi-darkness. The case enclosing the movement of the watch handed over is a standard case; the final model will be equipped with an extra-thick toughened glass and an extra-thick stainless-steel back."[16]

Over the months that followed, it is fascinating to see, one by one, the now familiar characteristics of the Submariner emerge in the letters going back and forth between Rolex and the Royal Navy. In October 1954, Winter dispatched what he describes as a "new model of the Rolex Oyster Perpetual Type 6150"[17] to Gutteridge for testing in the Mediterranean.

The Ref. 6150 "was manufactured in collaboration with the Admiralty."[18] It was, wrote Winter to Gutteridge, "fitted with a rotating bezel and a specially designed winding crown, toughened glass and back which permits this watch to be used with complete safety underwater down to a depth of at least 600 feet."[19]

He continued to explain "that the dial diameter is slightly larger than that of the existing model 6150 and, of course, we think there is a tremendous advantage in the rotating bezel insofar as just before the diver goes over the side, he can set the Zero Triangle with its radium point opposite the minute hand of his watch."[20]

In August 1955, three-and-a-half years after the initial dives in Loch Fyne, the contract with the Royal Navy was finally in place and a press release was drafted and approved by the Admiralty.

ROYAL NAVAL DIVERS NOW HAVE THE TIME UNDERWATER

Royal Naval divers have been issued officially with a new type of wrist-watch, developed by Hans Wilsdorf, 74-year-old inventor of the Oyster waterproof wrist-watch and the Perpetual self-winding wrist-watch, and Chairman of The Rolex Watch Company Limited.

The watch is self-wound by a rotor mechanism as the diver walks or swims. The dial is a jet-black instrument type illuminated for extra clear reading underwater and at night. This official Royal Naval divers' watch was put into production only after two years of exhaustive Admiralty exposure and timing tests in England and the Mediterranean. It has proved itself waterproof even down to 122 metres (400 feet) below the surface of the water.

Prolific inventor Dimitri Rebikoff revolutionized underwater photography with the development of an underwater lighting system that revealed the vivid colours of the aquatic world.

A DIVER'S DREAM COME TRUE

Dear René,
San Remo 12.1.1953
Your letter Re Mr Rebikoff is very interesting, and it seems to open fresh possibilities for Rolex becoming well known for our waterproof watches. The more we concentrate upon same the better for our future.[1]

In his letter to Jeanneret, Wilsdorf also ruminated over various names: "I like the name Deep Sea Special better than Frogman. I am sure that Nautilus is already registered."[2]

Military divers were far from being the only human visitors to the world under the waves. Wilsdorf was now a man in his seventies. Nevertheless, he was as engaged as ever with the latest advances in technology, determined to create ever better watches to meet the needs of a rapidly developing world, and one of the most startling developments of the 1950s was the explosion in underwater exploration. Although the Submariner served the needs of the armed forces, its potential for civilian diving was immense, as Rolex realized from the outset.

Men had been diving beneath the waves for millennia searching for pearls or sponges, and by the beginning of the 16th century, illustrations of primitive diving apparatuses began to appear in printed texts. In the early 19th century, a template was established that would alter little for well over a century: a bronze or copper spherical helmet, attached by a flexible collar to an airtight suit, fed with air delivered by a pipe connected to a pump at the surface. One could either enjoy the mobility and relative freedom of swimming underwater for as long as the lungs permitted, or achieve greater depths for longer periods of time in a cumbersome diving suit.

As this technology improved, greater depths were achieved. Underwater exploration was being revolutionized by a French naval officer with an aquiline nose and an invention called the Aqualung, which fed the diver with breathable air stored under pressure in tanks strapped to the back. In 1943, Jacques-Yves Cousteau and his collaborator Émile Gagnan had tested a prototype of this novel apparatus in the Marne River; 10 years later, the charismatic Cousteau and his invention were world-famous.

As early as 1865, divers had been experimenting with self-contained diving equipment fed by compressed air stored in portable tanks strapped to their bodies. Over the years, various incremental improvements had been made, while useful accessories including flippers, facemasks, spearguns, pressure gauges, and so forth had been developed. The breakthrough offered by the Cousteau-Gagnan patent allowed "the diver to breathe comfortably irrespective of his position in relation to the vertical, and at the same time ensures the completely automatic functioning of the apparatus."[3]

The freedom this system offered the diver was nothing short of magically transformative, as Cousteau wrote in an article for *National Geographic* in October 1952.

"The Aqualung frees a man to glide, unhurried and unharmed, fathoms deep beneath the sea. It permits him to skim face down through the water, roll over, or loll on his side, propelled along by flippered feet.

"No cables or hoses connect the aqualunger to the upper world. No heavy armour weights him down. Tanks strapped upon his back feed him compressed air in amounts carefully regulated to equalise the pressure within his body to the pressure of the sea without. In shallow water or in deep, he feels its weight upon him no more than do the fish that flicker shyly past him."[4]

Dimitri Rebikoff, to whom Wilsdorf referred in his letter, was another pioneer in the underwater world. Born to Belarusian parents in 1921 (his father had been an admiral in the Tsarist navy), fluent in Russian, English, German and French, Rebikoff was an engineering prodigy. At age 11, he started his own camera repair business, and after studying engineering and physics at the Sorbonne, he became a radio engineer. Taken prisoner by the Nazis in 1943, Rebikoff was put to work repairing radios in Munich, and with the end of the fighting, he purloined a camera and set off for Paris where he became known for colour photography, which was beginning to become popular. A talented and prolific inventor, Rebikoff soon patented his first electronic flashbulb. Other inventions followed and he established a company in Lausanne. By the late 1940s, Rebikoff was an acknowledged authority on photography, and wrote for an issue of *Science et Vie* in which there was also an article about underwater photography by members of the Club Alpin Sous-Marin (CASM), an organization formed by enthusiasts for the fledgling sport of scuba diving.*

"I wondered what this mysterious Club Alpin Sous-Marin could be," he mused. "It had been described to me in Paris as one of those snob groups which infest the Côte d'Azur."[5] Curious, Rebikoff decided to stop off in Cannes, where the club was based, on his way to vacation in Corsica with his fiancée, and look up Henri Broussard, the man who had founded the club in 1946.

A gruff, eagle-nosed man in his mid-forties, Broussard ran a garage in Cannes, where he kept the compressor that filled the club's air tanks. He greeted Rebikoff with the words: "You're diving tomorrow at eleven-thirty sharp."[6]

"The next day we went to the Palais des Sports in front of the Bay of Cannes. It was a war ruin in those days," recalled his future wife, Ada, looking back on that long-distant summer half a century later. "They put the *'scaphandre'* (an aqualung, in this instance) on Dimitri. His size-14 feet were too big for their fins. Then Dimitri got the order, 'You see the rock there, swim to it.' The water was about 3 to 4 metres deep. Dimitri was not a good swimmer. He pulled himself along the bottom with his hands."[7]

Inelegant though they may have been, those first few minutes of pulling himself along the seabed would change his life. Within three days, he was diving at a depth of 40 metres (131 feet). To use a marine metaphor, he was hooked. He and Ada never made it to Corsica; instead they spent a fortnight in Cannes diving with Broussard, whom the couple found to be "an extremely sensitive and generous man under his reserved and boorish exterior".[8] Rebikoff soon moved to Cannes, where he and Ada married in 1952, to develop the nascent field of underwater photography.

Inspired and enraptured by what he had encountered beneath the waves, he worked quickly and by 1950, he had perfected an electric flashbulb for underwater use as well as watertight housings for cameras. Today, we take high-quality underwater film and photographs for granted, but in the early 1950s, the subaquatic world had yet to be appreciated in all its polychromatic splendour. "Previously, all attempts at photography in colour at great depth had failed owing to lack of sufficient light, and we knew nothing of the depths except blues and greens and, occasionally, yellows a few metres down," observed Rebikoff. But even this experienced photographer and inventor had not imagined the visual riches that would be unlocked by his underwater lighting system. "That powerful source of artificial light, the electronic flash, which has been used for a long time by reporters, brought us a tremendous surprise and upset all our well-established ideas: the great depths are not blue, but yellow, orange, and red with a multiplicity of shades which cannot be found on land."[9]

* Scuba is an acronym for "self-contained underwater breathing apparatus".

Dimitri Rebikoff in the early 1950s.

"IN A DIVE, IT IS NOT THE TIME OF DAY THAT COUNTS BUT THE NUMBER OF MINUTES THAT HAVE ELAPSED SINCE ENTERING THE WATER."

DIMITRI REBIKOFF

It was as if a curtain had suddenly been pulled back. An entirely new world of spectacularly vivid colour had been opened up in – excuse the pun – a flash. Recording their subaqueous adventures in order to be able to share their discoveries with a wider audience became one of the chief goals of the early scuba divers, who were keen to broaden the appeal of their sport and to demonstrate its value in terms of scientific research and exploration. Rebikoff made some of the earliest underwater films, and with the equipment he developed, he was one of the founding fathers of underwater filmmaking.

In the late 1940s and early 1950s, the pace of, and enthusiasm for, underwater exploration off the Côte d'Azur exploded. Within a couple of years, Rebikoff had become a prominent figure in the fast-developing and exciting world of scuba diving, his fame growing with the publication of his first book *L'Exploration Sous-marine* in 1952 and the invention of a one-man underwater vehicle called *Pegasus*, described as "a cross between a torpedo and an airplane."[10]

In those carefree summers after the war, scuba diving became the pastime of playboys and thrill-seeking aristocrats in search of a summertime adrenaline rush outside of the skiing, shooting and hunting seasons. As well as finding itself at the forefront of discovery and innovation, the world of aqualung diving was glamorous and exotic.

The scuba-diving community on the Côte d'Azur in those days was close-knit, and among the explorers, former naval officers, underwater cameramen, and devil-may-care playboys was Rolex's René-Paul Jeanneret, a keen aqualunger. At the beginning of 1953, he had been elected a "corresponding" member of the Institut de Recherches Sous-Marines (IRSM), in which Rebikoff was also active.

A version of the Submariner, Ref. 6200, 38 mm, dating from 1954, with numerals 3, 6 and 9 on its dial – a feature usually found on the Explorer – fitted with an expandable bracelet.

Opposite: Underwater pioneer Dimitri Rebikoff on a dive, early 1950s.

RENÉ-PAUL JEANNERET BROUGHT TO ROLEX HIS OWN EMPIRICAL UNDERSTANDING OF WHAT WAS REQUIRED OF SUCH A TIMEPIECE FOR DIVERS.

In early March of the same year, upon his return from a business trip to the U.S., Jeanneret replied with his thanks, asking what fees were due as a corresponding member of the IRSM and enclosing his signed application form for membership in the CASM.

His reply as it is preserved on a flimsy pink carbon copy in the Rolex Archives offers a remarkable, privileged insight into not so much the birth, not even the gestation, but the very conception of one of the most famous watches ever made.

"Furthermore, I am pleased to inform you that in my absence a prototype was made based on the drawing that you approved in January," he wrote. "However, we have thought of another couple of refinements of details that we are working on now. As soon as the final prototype has been developed, we will send it to you so that it can be submitted to your practical tests."[11]

Their correspondence also contains personal observations. At one point, Jeanneret writes that his dream would be to bring a Greek amphora up from the seabed off the coast of Hyères Islands and asks to join Rebikoff, should he return to dive on the site where the amphorae were found. The seabed of the Mediterranean was littered with thousands of years of shipwrecks that had lain undisturbed for millennia until being eagerly explored by those early aqualungers of the 1950s, and the ancient clay storage jars were popular souvenirs. Once discovered, the preferred method of bringing them to the surface was to empty them of sand and detritus, fill them with compressed air and send them floating upwards. On their way to the surface, they would gain enough speed to shoot out of the water like some 2,000-year-old clay missile fired from a submarine. By September 1954, Jeanneret was the proud owner of a Sestius amphora*, which was "admired by all my friends who have had the opportunity to see it at my house."[12]

As an enthusiastic amateur diver, Jeanneret brought to Rolex his own empirical understanding of what was required of such a timepiece for divers. And, importantly, he had personal access to the highest-profile and most experienced divers of the time, chief among them Rebikoff, who contributed invaluable practical underwater experience as well as his unique insight as a prolific inventor.

Rebikoff specifically mentioned his work with Rolex in his celebrated book *Free Diving*, which was partly an account of his own experiences underwater and partly a manual and

* The Sestius family was prominent in politics and commerce at the end of the Republic and beginning of the Roman Empire and possessed an amphora factory near modern day Porto Ercole.

well-informed introduction to all aspects of scuba diving. "The watertight watch made by Rolex has a reinforced case which allows it to withstand a depth of 100 metres. It is very important for determining the duration of dive, varying with depth, according to the table or the curve of dive." He found the 60-minute bezel particularly useful: "The reason for this is that in a dive, it is not the time of day that counts but the number of minutes that have elapsed since entering the water."[13]

On 7 April 1953, he wrote to Jeanneret to say "the prototype of the waterproof diving watch appears particularly successful and will be put to the test as soon as possible."[14]

But in the meantime, he had some constructive criticism: "The red triangle will probably already become invisible below 10 metres," wrote Rebikoff, suggesting instead what would become one of the defining features of the bezel. "It would be better to replace it with a white triangle."[15] A few days later, he received Jeanneret's reply: "We are currently studying modifying the bezel so as to make the triangle snow white rather than red."[16]

There were occasional growing pains along the way. For instance, Rebikoff returned the prototype "because its operation becomes quite irregular. Indeed, it stops after about 6 hours of rest during the night, even on the train."[17]

Jeanneret was at a loss as to account for the stoppages, explaining that the watch was fitted with "an exceptionally sensitive automatic movement guaranteeing full rewinding after only a few hours of wear,"[18] but promised that he would get to the bottom of the mystery once he had received the watch. In all likelihood, Rebikoff was unfamiliar (as the Royal Navy divers on Loch Fyne had been) with the experience of wearing a self-winding watch, relying solely on his movements rather than winding it fully before wearing. But whatever the reason, it was of relative unimportance, with Rebikoff magnanimously describing it as "certainly a minor problem, and I would be happy to be able to continue testing as soon as it is fixed."[19]

Back and forth went the letters until, on 28 May, Jeanneret wrote with the momentous news that Rolex was "going to proceed with the manufacture of a pilot series of this model, a series which will of course be equipped with a white triangle on the bezel."[20]

Earlier that day, at a technical meeting of the board of Rolex, the criteria for a pilot series of 100 watches had been established: a 38 mm case supplied by Vallon with a reinforced back; a 9¾ ligne Perpetual movement; Nivaflex spring; black Royal Navy type dial with radium indices; Twinlock crown; "tropic" glass resistant to 200 metres (656 feet); and an extendable steel bracelet.

One of Rebikoff's observations had been that the glass was prone to scratching. "We do not see how we could prevent the plexiglass from being scratched when diving, except by means of a grille which, however, would seriously impair visibility," conceded Jeanneret. But he reassured Rebikoff that "this new model will be fitted with a glass 2 mm thick which, once scratched, can be repolished without difficulty by a competent watchmaker."[21]

"As for the moving bezel, it is obvious that it will prove invaluable for many uses other than diving."[22]

Rebikoff was delighted: "I am very happy to hear that you are making a pilot series which of course will have to be tried systematically under the harshest working conditions, during IRSM research, as well as during the shooting of

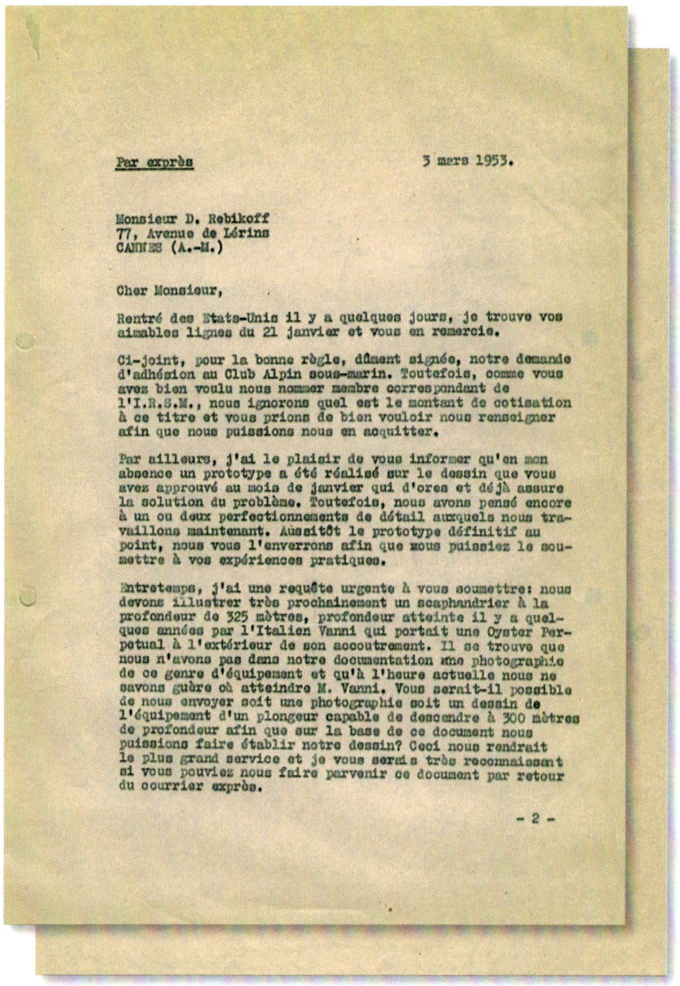

Par exprès — 3 mars 1953.

Monsieur D. Rebikoff
77, Avenue de Lérins
CANNES (A.-M.)

Cher Monsieur,

Rentré des Etats-Unis il y a quelques jours, je trouve vos aimables lignes du 21 janvier et vous en remercie.

Ci-joint, pour la bonne règle, dûment signée, notre demande d'adhésion au Club Alpin sous-marin. Toutefois, comme vous avez bien voulu nous nommer membre correspondant de l'I.R.S.M., nous ignorons quel est le montant de cotisation à ce titre et vous prions de bien vouloir nous renseigner afin que nous puissions nous en acquitter.

Par ailleurs, j'ai le plaisir de vous informer qu'en mon absence un prototype a été réalisé sur le dessin que vous avez approuvé au mois de janvier qui d'ores et déjà assure la solution du problème. Toutefois, nous avons pensé encore à un ou deux perfectionnements de détail auxquels nous travaillons maintenant. Aussitôt le prototype définitif au point, nous vous l'enverrons afin que nous puissiez le soumettre à vos expériences pratiques.

Entretemps, j'ai une requête urgente à vous soumettre: nous devons illustrer très prochainement un scaphandrier à la profondeur de 325 mètres, profondeur atteinte il y a quelques années par l'Italien Vanni qui portait une Oyster Perpetual à l'extérieur de son accoutrement. Il se trouve que nous n'avons pas dans notre documentation une photographie de ce genre d'équipement et qu'à l'heure actuelle nous ne savons guère où atteindre M. Vanni. Vous serait-il possible de nous envoyer soit une photographie soit un dessin de l'équipement d'un plongeur capable de descendre à 300 mètres de profondeur afin que sur la base de ce document nous puissions faire établir notre dessin? Ceci nous rendrait le plus grand service et je vous serais très reconnaissant si vous pouviez nous faire parvenir ce document par retour du courrier exprès.

- 2 -

Letter from René-Paul Jeanneret to Dimitri Rebikoff, 3 March 1953.

Above and opposite: A selection of advertisements, handbooks and endorsements for the Submariner.

three underwater films in colour that I have to do this summer. Of course, I will do my best to include the watches in the foreground when shooting."[23]

It is thrilling to see the watch develop and take shape in the correspondence between the two men in the spring of 1953. Today, we view the Submariner with the certainty of history, but at the time of its invention the very concept of a diving watch was alien, unheard of. And yet, as the spring of 1953 turned into summer, there was the growing confidence that a promising future awaited what was known internally as the "*pièce* Rebikoff". Although the watch would soon acquire another name as the minutes of the 28 May technical meeting record.

"Mr Jean HUGUENIN points out that these pieces should be sealed with a rubber gasket because the metal gaskets will not resist the pressure at 200 metres." And it was decided that "this piece will carry the name: 'SUBMARINER'."[24]

From June until October, Rebikoff and his IRSM team tested the Submariner exhaustively. In all, 132 dives were undertaken at depths between 12 and 60 metres (39 to 197 feet) in waters ranging from a chilly 12 degrees Celsius to an almost tropical 28.

"The watch was subject to no special precautions during the dives," reported Rebikoff. "On the contrary, in the second month of the trials, it was dropped during loading operations from a height of about six feet onto the cement quay. Despite this very violent shock, the accuracy of the watch was not affected in the least. The only visible damage was a flake of luminous material falling from the minute hand."[25]

Rebikoff demonstrated an engineer's thoroughness in his five months of testing and seemed unable to find fault. "In spite of the high degree of salt in Mediterranean waters, combined between the dives with tropical temperatures and humidity, no trace of corrosion could be detected ... no trace of humidity could be found in the watch," notwithstanding that "several dives were performed with the winding crown pulled out, in the position for hand-setting."[26]

In a final bid to destroy the watch, Rebikoff attached it to a length of steel wire and dropped it overboard "to a depth of 120 metres, double the maximum to which we are able to dive with independent compressed air equipment. After an hour of this test, no water had infiltrated into the watch."

SCUBA DIVING BECAME THE PASTIME OF PLAYBOYS AND THRILL-SEEKING ARISTOCRATS IN SEARCH OF A SUMMERTIME ADRENALINE RUSH.

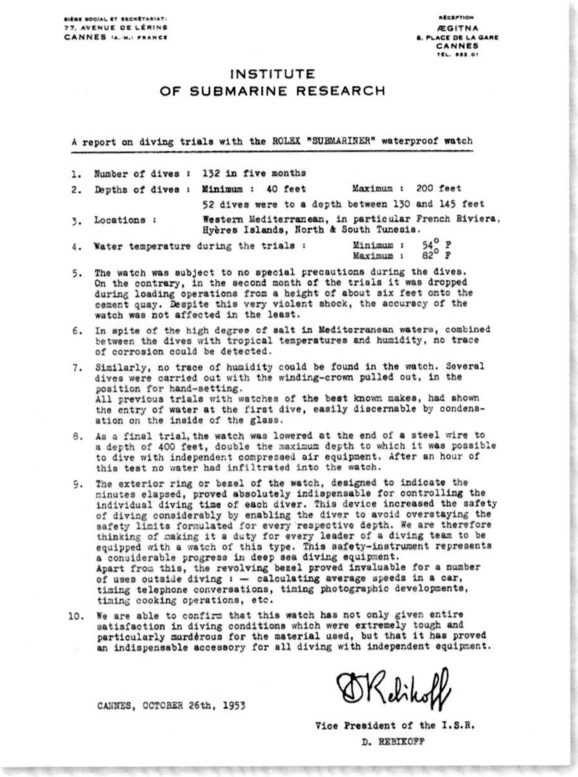

SIÈGE SOCIAL ET SECRÉTARIAT:
77, AVENUE DE LÉRINS
CANNES (A.-M.) FRANCE

RÉCEPTION
ÆGITNA
8, PLACE DE LA GARE
CANNES
TEL. [illegible]

INSTITUTE
OF SUBMARINE RESEARCH

A report on diving trials with the ROLEX "SUBMARINER" waterproof watch

1. Number of dives : 132 in five months
2. Depths of dives : Minimum : 40 feet Maximum : 200 feet
   52 dives were to a depth between 130 and 145 feet
3. Locations : Western Mediterranean, in particular French Riviera, Hyères Islands, North & South Tunesia.
4. Water temperature during the trials : Minimum : 54° F Maximum : 82° F
5. The watch was subject to no special precautions during the dives. On the contrary, in the second month of the trials it was dropped during loading operations from a height of about six feet onto the cement quay. Despite this very violent shock, the accuracy of the watch was not affected in the least.
6. In spite of the high degree of salt in Mediterranean waters, combined between the dives with tropical temperatures and humidity, no trace of corrosion could be detected.
7. Similarly, no trace of humidity could be found in the watch. Several dives were carried out with the winding-crown pulled out, in the position for hand-setting.
   All previous trials with watches of the best known makes, had shown the entry of water at the first dive, easily discernable by condensation on the inside of the glass.
8. As a final trial, the watch was lowered at the end of a steel wire to a depth of 400 feet, double the maximum depth to which it was possible to dive with independent compressed air equipment. After an hour of this test no water had infiltrated into the watch.
9. The exterior ring or bezel of the watch, designed to indicate the minutes elapsed, proved absolutely indispensable for controlling the individual diving time of each diver. This device increased the safety of diving considerably by enabling the diver to avoid overstaying the safety limits formulated for every respective depth. We are therefore thinking of making it a duty for every leader of a diving team to be equipped with a watch of this type. This safety-instrument represents a considerable progress in deep sea diving equipment.
   Apart from this, the revolving bezel proved invaluable for a number of uses outside diving : — calculating average speeds in a car, timing telephone conversations, timing photographic developments, timing cooking operations, etc.
10. We are able to confirm that this watch has not only given entire satisfaction in diving conditions which were extremely tough and particularly murdèrous for the material used, but that it has proved an indispensable accessory for all diving with independent equipment.

CANNES, OCTOBER 26th, 1953

D Rebikoff

Vice President of the I.S.R.
D. REBIKOFF

He could only report that the watch had "given entire satisfaction in diving conditions which were extremely tough" and "proved an indispensable accessory for all diving with independent equipment."[27]

By contrast, "all previous trials with watches of the best known makes had shown the entry of water from the first dive, easily recognizable by the accumulation of condensation on the inside of the glass."[28]

But more than outperforming its rivals, surviving beyond conditions that would have killed anyone wearing it and generally overfulfilling its functions, one thing in particular impressed Rebikoff: the rotating bezel with its white triangle, which allowed the diver to time his or her stay underwater. Up until then, scuba divers had been made aware of the passage of time by various methods including firing a gun into the water to alert divers that they needed to return to the surface.

The rotating bezel had first been tried at Rolex on the 1938 Zerographe, but it was clearly still a novelty in 1953, as Rebikoff's sense of wonder betrays. "The exterior disc of the watch, designed to indicate the minutes elapsed, showed itself to be absolutely indispensable for controlling the individual diving time of each diver. This mechanism showed itself to increase the safety of diving considerably by enabling the diver to avoid exceeding the time limits fixed by tables for every depth reached. We are therefore thinking of making the wearing of a watch of this type mandatory for at least the leaders of a diving team. This safety device represents considerable progress in deep-sea diving equipment. Apart from this, the revolving disc proved very valuable for a number of uses outside diving: calculating average journey times by car, timing telephone conversations, timing photographic developments, timing cooking, etc."[29]

When Wilsdorf wrote from San Remo in January 1953, he added as a PS: "A 'deep-sea' model will surely find a big market everywhere."[30] He could hardly have dared imagine how big and enduring that market would become.

A diving report issued by the Institute of Submarine Research and signed by Dimitri Rebikoff, 1953.

Opposite, from left: Hans Wilsdorf with Werner Ryser, Walter Luthy and René-Paul Jeanneret in the year Hans Wilsdorf celebrated four anniversaries: his 70th birthday, 50 years of activity in the watchmaking world, the 25th birthday of the Oyster and the 20th birthday of the Perpetual rotor, 1951.

Divers safely time their decompression using Submariners in an expedition off the Tunisian coast in 1955.

AT THE CENTRE OF THE WORLD OF DIVING

The Submariner would eventually become famous, promoted in advertising and frequently seen on famous wrists and in films, but in the early 1950s, all of that lay in an unknowable future. And while the concept of a "professional" or "tool" watch had yet to imprint itself on the watch market, the Submariner was not entirely without precedent. There had been waterproof watches before. As early as 1922, Rolex had registered the name "Submarine" for a watch housed within a double-case system.

Cased as normal, the watch was then attached by a hinge inside a second case that was screwed shut. When the wearer wanted to wind or set the watch, the outer case had to be unscrewed before the winding crown could be accessed. Complex and inconvenient though it may have been, it was nevertheless the best that could be done in 1922. This clumsy system was soon replaced by Rolex's screwed-down winding crown, which maintained a perfect seal.

By the early 1950s, however, the pace of development was accelerating, and along with tests by the Royal Navy and Rebikoff, there is an abundance of correspondence in the Rolex Archives confirming that many other civilian bodies were also assessing the fitness of Rolex watches.

One typical example is the letter Jeanneret received at the beginning of November 1952 from Milan: "I am addressing you on behalf of the National Undersea Expedition to ask for your collaboration with an enterprise that will undoubtedly be interesting and advantageous for you," began his correspondent, Dr Alberto Grazioli, who was writing to inform him of an expedition that would "leave for the Dahlak Islands (Red Sea) on 23 November" and would "remain there for three months.

"This is the second such expedition in the world and the first Italian one.

"Its goals are: scientific, documentary and sporting.

"Since our activity will be mainly underwater, we turn to you to ask if you would agree to make available to us, for free, some of your waterproof watches with which other sporting expeditions have already been equipped.

"We would need six of these watches."

In return, the expedition was offering Rolex the opportunity to advertise its involvement with the expedition: "We give you our formal consent to declare to the public that 'the National Undersea Expedition has chosen for its undersea activity waterproof watches from the house Rolex'.

"We will give you a detailed report on the behaviour of these watches on our return.

"The watches would be returned to you for you to examine if this is of interest to you."

SUBMARINE
ROLEX

The good doctor also offered to supply photographs of the expedition, and said that any films of the expedition "would show that underwater explorers did indeed wear waterproof watches during their dives."[1]

Jeanneret felt the opportunity was not to be missed and prepared a detailed memo to Marie Meinen, Head of Production, outlining the expedition's requirements of six "Rolex Oyster perpetual steel watches with 10.5 ligne movements, expanding bracelets and luminous dials." While the watches were to have standard case backs and glasses, "particular care should be taken concerning their water-resistance as this concerns an undersea expedition."[2] A later memo in the Rolex Archives suggests that at least four of the watches supplied were the Ref. 6150.[3]

Gentlemen,

I have the pleasure to express to you my complete satisfaction with the operation and behaviour underwater of your Rolex Oyster Perpetual chronometer, worn by me during the national submarine expedition in the Red Sea.

Setting the time against the time signal on my departure from Italy, I noticed on my return that the watch had not changed, although I had never set the time or rewound it.

[Worn for several months in what the doctor described as the "torrid climate of the Red Sea region", the watches had been repeatedly "carried out in the extremely salty water of the Red Sea" for long deep dives.]

The only trace of the hard tests to which the watch has been subjected is found on the unbreakable crystal, which has scratches, scratches which are due to contact with madrepores (which have also left many marks on my skin).

Please accept, Gentlemen, the expression of my highest consideration.

Dr Luigi Stuart Tovini[4]

As Tovini's observations demonstrate, such a watch was regarded and evaluated as a functional item of diving equipment rather than a wristwatch as it is understood in the 21st century. In the early 1950s, the idea that such an instrument could be worn in civilian life would have been considered almost as outlandish as wearing a diving suit on city streets. It had been scarcely more than two decades since production of wristwatches had overtaken that of pocket watches, and the prevailing taste in men's wristwatches at the beginning of the 1950s was for slim timepieces that barely disturbed the cuff of a gentleman's shirt.

Rolex's first American manager recalled in later life how he had encountered resistance in the market to a comparatively thick steel watch on a steel bracelet at a time when the popular idea of a man's wristwatch was something the size of a coin, in gold, on a leather or alligator strap: "I used to travel with my car and my small collection, going all around the country, and people used to say to me, 'What do you want me to do with your onions?'"[5]

To 21st-century eyes, the Rolex Oysters of the postwar years seem small, if anything; however, at the time they were considered large. Moreover, the Oyster's capacity to resist conditions that would kill the unprotected wearer had already been established: the successful 1953 attempt to scale Mount Everest by a team equipped with Rolex watches was an example. In this context, it is easy to imagine how it could be argued that if a standard production model Rolex Oyster could withstand such an extreme environment, then the need for a discrete diving model would be restricted to a highly specialized market. Certainly, there was enthusiasm for the watch that would become the Submariner at the very highest levels of Rolex management, viz., Wilsdorf's letter to Jeanneret, but the archives indicate that it was Jeanneret's own personal interest in diving and his curiosity for new technology and avant-garde design that gave the project much of its early impetus.

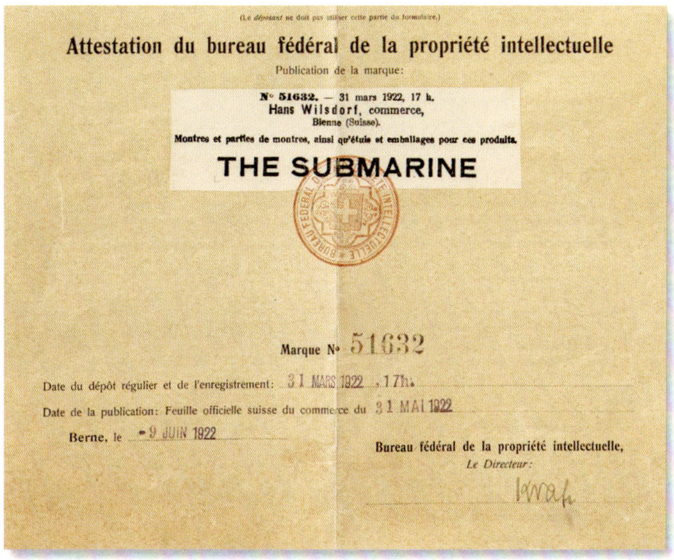

(Le déposant ne doit pas utiliser cette partie du formulaire.)

Attestation du bureau fédéral de la propriété intellectuelle

Publication de la marque:

N° 51632. — 31 mars 1922, 17 h.
Hans Wilsdorf, commerce,
Bienne (Suisse).

Montres et parties de montres, ainsi qu'étuis et emballages pour ces produits.

THE SUBMARINE

Marque N° 51632

Date du dépôt régulier et de l'enregistrement: 31 MARS 1922, 17h.

Date de la publication: Feuille officielle suisse du commerce du 31 MAI 1922

Berne, le 9 JUIN 1922

Bureau fédéral de la propriété intellectuelle,
Le Directeur:

The certificate of registration for the intellectual property of the name "The Submarine", 1922.

Opposite: Presented in 1922, the Submarine marked the first step in Hans Wilsdorf's efforts to create a completely sealed watch case.

As far as Rolex at the Basel Fair of 1954 was concerned, the Submariner was still a niche product. Instead, the focus that year was the Turn-O-Graph Ref. 6202, which had been designed around the concept of the rotating bezel. The turning calibrated bezel (Patent No. CH312285, applied for in 1953, granted in 1956) has since become such a familiar sight on wristwatches that it is hard to imagine the excitement it generated at the time of its first appearance. The Turn-O-Graph was to showcase the versatility of this new feature and was presented as "the watch with a thousand and one uses."[6] The accompanying brochure advised how it could be used to remind the wearer of upcoming business meetings, or, rather more abstrusely, how it could be used to calculate the rate of production of a cigarette packing machine. By contrast, the Submariner Ref. 6204 was an addition to the model range and, except for the slightly bolder, less busily designed bezel-insert, was almost identical to the Turn-O-Graph; early models of the Submariner even used the same hands (the distinctive "Mercedes" hour hand was a later addition). However, the Submariner would quickly upstage its parent model, thanks in great part to Jeanneret's flair for marketing.

Jeanneret was compiling a rapidly growing dossier of endorsements from leading authorities on diving, and 1954 saw further encomia for the new model. In November of that year, Jeanneret received a detailed report from the Club d'Etudes Sous-Marines of Tunisia concerning 12 watches lent for use on an underwater archaeological expedition to the famous site of Mahdia just off the Tunisian coast.

"No precautions were taken to protect the watches,"[7] nevertheless they performed reliably and once again the rotating bezel was singled out for particular praise:

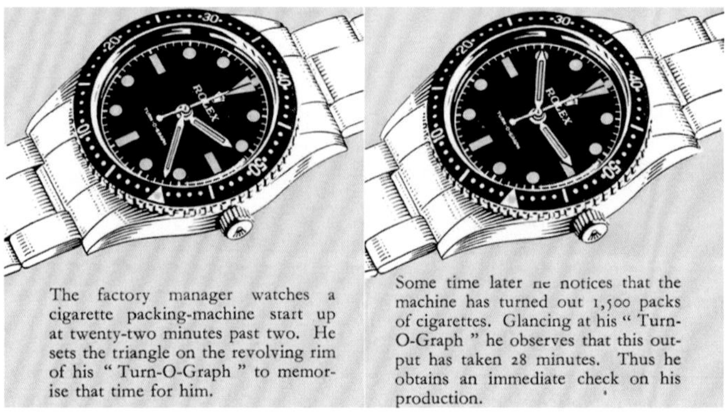

Extract from the Turn-O-Graph brochure, 1956.

Opposite: The first Turn-O-Graph, launched in 1953, was made of stainless steel and was 36 mm in diameter. This watch was designed around the concept of the rotating bezel.

"With the time indicator of the rotating bezel on the minute hand, each diver thus knew precisely his immersion time and when dive time reached 15 minutes, he would decompress at the desired stops," wrote the Club President on 15 November 1954.

"All the members of the Mahdia expedition consider that permanent control of dive times, thanks to the Rolex Submariner, was one of the essential factors of the safety system, which was applied, moreover, with success since we did not have the least decompression sickness to complain of.

"At the end of the expedition we were surprised to see:

– That all the watches showed the same time to within a few seconds.

– That neither the watches nor the bracelets showed the slightest trace of corrosion."[8]

As the popularity of diving grew, scuba clubs and associated organizations proliferated around the Mediterranean littoral, and in autumn of 1954, Jeanneret wrote to another of his diving buddies, a Marseille doctor and keen spearfisherman named Xavier Fructus, who had joined Cousteau's team as medical advisor in 1952 and by 1954 was Vice President of the Office Français de Recherches Sous-Marines (OFRS) in Marseille, to which Jeanneret presented three Submariners. The OFRS had come into existence in March 1953 as the successor to the earlier non-profit organization Campagnes Océanographiques Françaises (COF), which Cousteau had founded in 1951 following his resignation from the French navy. The OFRS was an engineering facility whose primary purpose was developing vehicles and instruments for use on Cousteau's expeditions, but it also undertook its own missions with its vessel, *L'Espadon*.

Fructus was just one of the names that appeared on an internal document compiled in October 1954 entitled "List of People to whom SUBMARINERS have been handed". Divided into seven categories – "For the purpose of being tested", "For under water expeditions", "Gifts", "Loans to diving clubs, etc.", "Direct sales from Geneva (amateur divers)", "For commercial purposes (distribution)", and "Miscellaneous" – this list demonstrated the range and efficacy of Jeanneret's strategy of placing the Submariner at the centre of the fast-growing world of diving in the 1950s.

ROLEX
TURN-O-GRAPH

Submariner, 1955, Ref. 6536, 38 mm, stainless steel, waterproof to 100 metres (330 feet).

Opposite: Divers working on a shipwreck off the Tunisian coast, 1955.

The Submariner once sold with an anchor, indicating the maximum depth of the watch (1979).
Opposite: Extracts from Submariner brochure, 1966.

The recipients included agents for diving equipment, such as Hamburg's Barakuda Gesellschaft für Wassersport and La Spirotechnique in Paris; diving instructors and professional divers; explorers and marine archaeologists; ship commanders; underwater cameramen and photographers; the Royal Canadian Navy, the Admiralty in London, and the Department of the Navy in Washington; as well as assorted figures of international military significance.

The start of 1955 brought with it a further glowing report from another diving club – this time from Henri Broussard, founder of the venerable CASM. His letter contained the now-familiar attestations to excellent performance of the watches over a number of dives and the usual rhapsodic description of the functionality of the rotating bezel. Clearly, the white triangle was doing its work well, and Broussard praised the legibility of the dial under the penumbral condition of the deeper dives, as well as during night dives.

"In summary, I would like to express my great satisfaction with this new Rolex model, which is destined to know the greatest success, and congratulate you on having developed a watch that also perfectly meets the needs of all those who practice water sports, hunting and most particularly underwater exploration."[9]

The only unusual thing about Broussard's letter was the watch to which he was referring: the "ROLEX OYSTER PERPETUAL SUB AQUA".[10]

The Oyster Perpetual Sub Aqua was not, in fact, a new watch, but rather, merely a new name for the Submariner. Originally, Broussard had written extolling the benefits of the Submariner only to be informed by Jeanneret that he had "very recently been warned that the brand name Submariner had been registered by a watch company in the United States which prohibits us from using it in that country. Given the importance of this market, we are obliged to stop calling our Submariner by that name, and we have decided to opt for the name 'Sub-Aqua'."

Obligingly, Broussard changed the name and submitted his report anew, but the clash of nomenclature seems to have been resolved relatively swiftly as it only appears fleetingly in the archives, and the number of watches bearing this name on their dials is correspondingly small.

By the second half of the 1950s, Rolex's Submariner had proved itself invaluable to the community of professional divers and was ready to make an impression on the wider world.

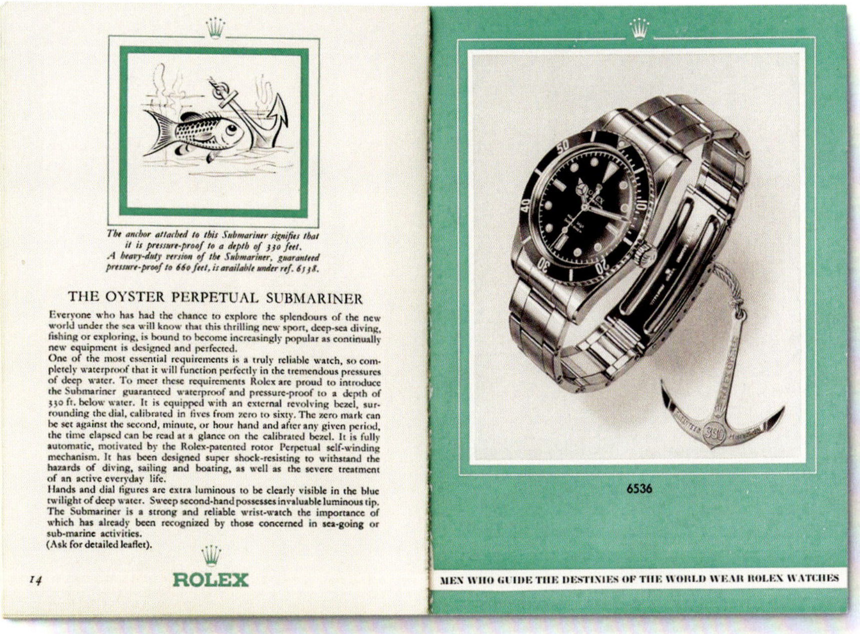

The anchor attached to this Submariner signifies that it is pressure-proof to a depth of 330 feet. A heavy-duty version of the Submariner, guaranteed pressure-proof to 660 feet, is available under ref. 6538.

THE OYSTER PERPETUAL SUBMARINER

Everyone who has had the chance to explore the splendours of the new world under the sea will know that this thrilling new sport, deep-sea diving, fishing or exploring, is bound to become increasingly popular as continually new equipment is designed and perfected.
One of the most essential requirements is a truly reliable watch, so completely waterproof that it will function perfectly in the tremendous pressures of deep water. To meet these requirements Rolex are proud to introduce the Submariner guaranteed waterproof and pressure-proof to a depth of 330 ft. below water. It is equipped with an external revolving bezel, surrounding the dial, calibrated in fives from zero to sixty. The zero mark can be set against the second, minute, or hour hand and after any given period, the time elapsed can be read at a glance on the calibrated bezel. It is fully automatic, motivated by the Rolex-patented rotor Perpetual self-winding mechanism. It has been designed super shock-resisting to withstand the hazards of diving, sailing and boating, as well as the severe treatment of an active everyday life.
Hands and dial figures are extra luminous to be clearly visible in the blue twilight of deep water. Sweep second-hand possesses invaluable luminous tip.
The Submariner is a strong and reliable wrist-watch the importance of which has already been recognized by those concerned in sea-going or sub-marine activities.
(Ask for detailed leaflet).

14 ROLEX

6536

MEN WHO GUIDE THE DESTINIES OF THE WORLD WEAR ROLEX WATCHES

BY 1960, ROLEX'S SUBMARINER HAD PROVED ITSELF INVALUABLE TO PROFESSIONAL DIVERS AND WAS READY TO MAKE AN IMPRESSION ON THE WIDER WORLD.

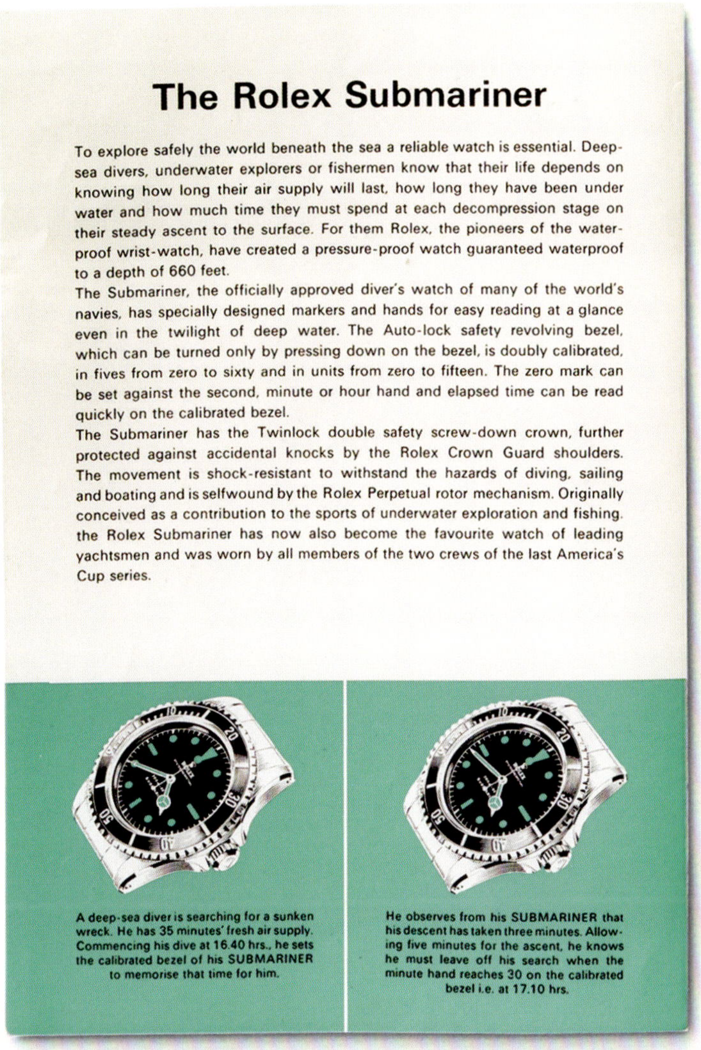

The Rolex Submariner

To explore safely the world beneath the sea a reliable watch is essential. Deep-sea divers, underwater explorers or fishermen know that their life depends on knowing how long their air supply will last, how long they have been under water and how much time they must spend at each decompression stage on their steady ascent to the surface. For them Rolex, the pioneers of the water-proof wrist-watch, have created a pressure-proof watch guaranteed waterproof to a depth of 660 feet.
The Submariner, the officially approved diver's watch of many of the world's navies, has specially designed markers and hands for easy reading at a glance even in the twilight of deep water. The Auto-lock safety revolving bezel, which can be turned only by pressing down on the bezel, is doubly calibrated, in fives from zero to sixty and in units from zero to fifteen. The zero mark can be set against the second, minute or hour hand and elapsed time can be read quickly on the calibrated bezel.
The Submariner has the Twinlock double safety screw-down crown, further protected against accidental knocks by the Rolex Crown Guard shoulders. The movement is shock-resistant to withstand the hazards of diving, sailing and boating and is selfwound by the Rolex Perpetual rotor mechanism. Originally conceived as a contribution to the sports of underwater exploration and fishing, the Rolex Submariner has now also become the favourite watch of leading yachtsmen and was worn by all members of the two crews of the last America's Cup series.

A deep-sea diver is searching for a sunken wreck. He has 35 minutes' fresh air supply. Commencing his dive at 16.40 hrs., he sets the calibrated bezel of his SUBMARINER to memorise that time for him.

He observes from his SUBMARINER that his descent has taken three minutes. Allowing five minutes for the ascent, he knows he must leave off his search when the minute hand reaches 30 on the calibrated bezel i.e. at 17.10 hrs.

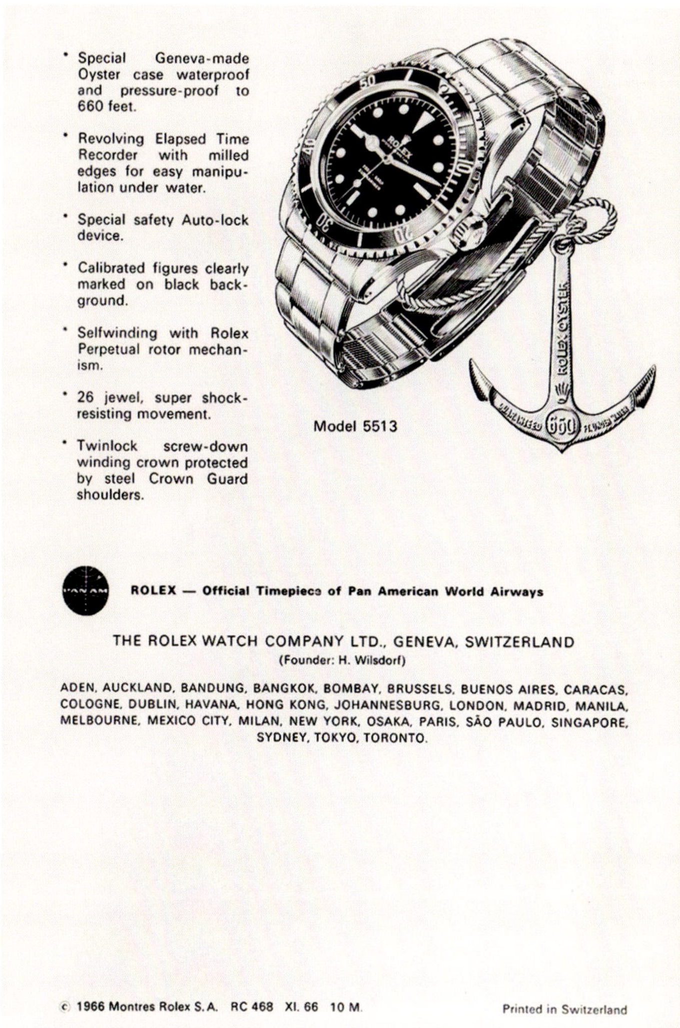

- Special Geneva-made Oyster case waterproof and pressure-proof to 660 feet.
- Revolving Elapsed Time Recorder with milled edges for easy manipulation under water.
- Special safety Auto-lock device.
- Calibrated figures clearly marked on black background.
- Selfwinding with Rolex Perpetual rotor mechanism.
- 26 jewel, super shock-resisting movement.
- Twinlock screw-down winding crown protected by steel Crown Guard shoulders.

Model 5513

ROLEX — Official Timepiece of Pan American World Airways

THE ROLEX WATCH COMPANY LTD., GENEVA, SWITZERLAND
(Founder: H. Wilsdorf)

ADEN, AUCKLAND, BANDUNG, BANGKOK, BOMBAY, BRUSSELS, BUENOS AIRES, CARACAS, COLOGNE, DUBLIN, HAVANA, HONG KONG, JOHANNESBURG, LONDON, MADRID, MANILA, MELBOURNE, MEXICO CITY, MILAN, NEW YORK, OSAKA, PARIS, SÃO PAULO, SINGAPORE, SYDNEY, TOKYO, TORONTO.

© 1966 Montres Rolex S.A. RC 468 XI. 66 10 M. Printed in Switzerland

A poster for the Academy Award-winning documentary
The Silent World by Jacques-Yves Cousteau and Louis Malle, 1956.

EXPLORING THE SILENT WORLD

"The world has become a bore."[1]

By the beginning of the 20th century, the store of locations to exploit and peoples to subjugate had been exhausted and European powers were reduced to claiming uninhabitable regions for the abstract prestige of being the first to plant their flags in barren land. Both poles had been reached by the outbreak of World War Two, and in 1953, humans finally reached the summit of Mount Everest, the highest point on Earth, and held aloft an ice axe from which fluttered the Union Jack – the flag of Great Britain. It seemed as if the world had been shorn of its last shreds of mystery.

"Our earth has grown very small in the mid-20th century. Even in the remotest parts of Africa or Polynesia, it has become impossible to avoid planes, radio sets and American cars; even the hinterland of Brazil and Greenland will soon have no secrets for us," lamented Dimitri Rebikoff.[2]

"All the pioneer types who, two or three centuries ago, would have sailed in quest of some Eldorado are today in despair. The world has become a bore. Once upon a time, a voyage was counted in years: nowadays, a few hours suffice. However, it sometimes takes years to procure the necessary visas and permits. Everywhere there are man-made frontiers: barbed wire, prohibitions and sardine tins are to be found even on 'inviolate' peaks. That is called progress."[3]

Rebikoff's tone was remarkably prescient and his remark about sardine tins betrays an awakening awareness of the damage humankind was inflicting on the planet.

But anyone who thought the wonders of the planet were exhausted just because even the polar regions and the highest mountains had been visited was, said Rebikoff, deeply mistaken: "Three-quarters of the surface of our earth is still virgin and unexplored. This marvellous country lies in the sea, hitherto known to us only by its surface, and seen for preference from a distance, from the deck of a liner or simply from the shore."[4]

Rebikoff's argument was that mankind had misunderstood the sea: seeing it as merely the aqueous interlude between land masses rather than the greater part of the planet. Familiarity, he said, had bred if not contempt, then complacency.

"The importance of the sea on our globe is apt to be forgotten. Nevertheless, it covers about 139 million square miles. Its greatest depths exceed 5,000 fathoms, the average depth being 2,000, whereas on land, Mount Everest alone reaches a height of 29,000 feet and the average height of the earth above sea-level is only some 300 feet."[5]

Rebikoff's eyes had been opened to the underwater world and he would do much to open the eyes of others. Increasingly widely available, relatively inexpensive scuba equipment was to bring this world within reach of anyone, to use his words, "well endowed with will-power and courage".[6]

More men and women came to think this way as, in the 1950s, diving captured the public imagination, and one man in particular came to embody this exotic and exciting undersea world.

DIVING CAPTURED THE PUBLIC IMAGINATION, AND ONE MAN IN PARTICULAR CAME TO EMBODY THIS EXOTIC AND EXCITING UNDERSEA WORLD: JACQUES-YVES COUSTEAU.

During the second half of the 20th century, Jacques-Yves Cousteau was not merely the world's most famous diver; he was one of its most famous men. Tireless, eloquent and charismatic, the sun of his reputation eclipsed all others, whether Yves Le Prieur, who can with some justification be called the father of modern skin diving, or Dimitri Rebikoff, who brought colour film to the underwater world. There were many divers, but there was only one Cousteau: his immense charm and ruthless ambition coincided with the explosion of colour magazines, colour films and television to create the first global environmental celebrity.

Beginning with a comparatively primitive film of 13 minutes in 1943 called *Par dix-huit mètres de fond*, his career as a film-maker would last 54 years, his final executive producer credit appearing in 1997, the year of his death. His films *The Silent World* and *World Without Sun*, made in 1956 and 1964 respectively, both won Academy Awards. His documentary series *The Undersea World of Jacques Cousteau* appeared on television from 1968 until 1976. Even his boat, the *Calypso*, was world-famous. In 1975, an eponymous record paying tribute to Cousteau and his research vessel was released by the American singer John Denver. "Calypso" climbed to number two in the Billboard Hot 100. With his toucan-like nose, sun-and-salt-fissured features, and his jolly, cherry-red knitted hat, his was one of the most recognizable faces of the second half of the 20th century. Appearing on countless talk shows and making innumerable public appearances, he held his audiences spellbound with his stories and mesmerized a world of armchair aquanauts with his polished films.

Jeanneret swiftly identified the potential publicity that he could bring to Rolex, and his quest for Cousteau is a persistent leitmotif of his correspondence in 1953 and 1954. His impatience to get a watch on Cousteau's wrist is palpable, leaping freshly from the letters held in the Rolex Archives. In August 1953, Jeanneret wrote to Rebikoff asking whether Cousteau would be in the Mediterranean in the coming weeks or whether he should send him a sample Submariner to Paris.[7] Rebikoff wrote back saying that Cousteau was in Greece, but that the *Calypso* was returning to Marseille, and he suggested that the best idea was to write to the *Calypso* in the old port of Marseille. But if he ever received Jeanneret's letter, he never replied.

Cousteau was proving elusive.

Opposite: Jacques-Yves Cousteau onboard his research vessel, the *Calypso* in 1983 *(top)*. The *Calypso* in Antarctica, 1973 *(bottom)*.

Following pages: ***The Undersea World of Jacques Cousteau***, a documentary series chronicling the undersea explorations of Jacques-Yves Cousteau and his crew aboard the ex-Royal Navy minesweeper *Calypso*, 24 October 1969.

However, Jeanneret was nothing if not tenacious. A year later, he was still trying to get Cousteau to endorse the watch and was attempting to contact him through Xavier Fructus. He wrote that he was willing to send watches "so that you yourself and Commander Cousteau would be able to use them in the course of your dives. If these pieces give as good results as those tried by other divers, notably by Mr Rebikoff, over the course of these last months, we would ask you if you would kindly give us some testimonials in the name of the Office of French Submarine Research, including, if possible, a testimonial signed by Commander Cousteau."[8]

Fructus replied saying "it would be a pleasure for the OFRS to supply you with these testimonials. I hope that we would also have the signature of Commander Cousteau, but I cannot promise you this at the moment because he is at sea with the *Calypso* and I have not yet been able to communicate the contents of your letter to him."[9]

Fructus clearly knew his boss well and was right not to make promises on his behalf. Cousteau is often mentioned anecdotally in connection with Rolex, but there is no record in the Rolex Archives of a direct connection or a formal relationship having been established at that time. However, flashes of what might be an early Submariner can be seen on Cousteau's wrist in his film *The Silent World*.

Ultimately, Jeanneret would be disappointed in his efforts to persuade Cousteau to give a formal report on the Submariner and to add Cousteau to his list of trophy endorsements. After Jeanneret left the company, Rolex continued its attempts to strike up a dialogue with Cousteau: in 1978, Walter Moser, then Press Officer at Rolex, wrote to Professor Henri Schwamm, Economic Adviser to the European Oceanic Association, of which Cousteau was Honorary President, with details of the recently inaugurated Rolex Awards for Enterprise. Schwamm said that he would certainly bring this information to Cousteau's attention.

It was not until September 1981 that Rolex would officially present Cousteau with a Submariner. By then, Jeanneret had been retired for seven years, and it was, for reasons that remain obscure, the Rolex affiliate in Mexico that hosted the reception at which the watch, which Rolex had been trying to place on the Commander's wrist for almost 30 years, was officially handed over. However, even though Cousteau had eluded him, Jeanneret could console himself with the fact that the younger generation of divers had no excuse for being ignorant of the Rolex Submariner, and that this awareness of Rolex among what could be called the "school" of Cousteau would pay rich dividends in the years to come.

One of the "students" at this "school" was Henri Delauze. The 23-year-old engineer had returned to France at the end of 1952, having completed his military service in the air force in Madagascar, which at the time was still a French colony. However, it was not in the skies above the island but under the waves of the Indian Ocean surrounding it that Delauze's future was determined as, over the three years he was stationed there, he became an enthusiastic diver. Born in Vaucluse and having spent some of his childhood in Toulon, he had studied in Aix-en-Provence, and it was to southern France that he returned once he had completed his military service.

A passionate diver, he gravitated to the bustling port city of Marseille, where he met diver Yves Girault, a member of Cousteau's team and another diving acquaintance of Jeanneret. Encouraged by Girault and his colleague Georges Beuchat, Delauze joined Cousteau and worked for the newly founded OFRS, making his first archaeological and speleological dives. But even as this constellation of divers was forming around the charismatic Cousteau, work was being undertaken that would eventually take diving to depths that Cousteau would have been incapable of imagining – work in which Rolex was to play an important part.

A lobby card for the film *The Silent World* by Jacques-Yves Cousteau and Louis Malle, 1956.

Opposite: Cousteau during the filming of *The Silent World*, 1956.

Sylvia Earle in the US Virgin Islands in 1970.

Right: A submersible in the Galápagos Islands in 2022 during an expedition by Sylvia Earle's Mission Blue.

Following pages: Sylvia Earle explores the seabed in a JIM suit, which she used in 1979 to set the world record for the deepest untethered walk at 381 metres (1,250 feet).

For almost a century, Rolex has accompanied explorers on expeditions to the highest mountains and deepest oceans – the perfect laboratory for testing its timepieces. Over the years, the company has moved from championing pure discovery to exploration for the sake of protecting the planet. Today, the Perpetual Planet Initiative reinforces Rolex's commitment to those using science to understand Earth and devise solutions to its environmental challenges.

ROLEX
Perpetual Planet Initiative

SYLVIA EARLE

Rolex Testimonee

"Her Deepness"[1], "First Lady of the Sea"[2], "Champion of the Deep"[3], "Hero from the Depths"[4]; these are just some of the informal titles that Sylvia Earle has been accorded, alongside more conventional prizes and honours, over the course of a lifetime devoted to oceanographic research and conservation.

A distinguished career in academia as a research fellow at Harvard was well under way when, in her mid-thirties, she joined the Tektite project, an all-female team of aquanauts living in an underwater laboratory and habitat off the island of Saint John in the Caribbean, in order to conduct innovative research there. It was while participating in the Tektite project that she began a relationship with Rolex lasting long into the 21st century.

"A watch was given to me as an aquanaut in 1970 as part of the Tektite II project that Rolex helped to equip. That watch became my favourite and most reliable 'dive buddy'. It was a beautiful, practical, stainless-steel instrument that enabled me to say how long I'd been down, how much longer I had to go; it was my lifesaving piece of equipment."[5]

She went below the waves a virtually unknown academic and surfaced to find herself famous, feted with: a ticker tape parade, a *Tonight Show* appearance, a White House celebration, a feature in *Life* magazine, the title of *Los Angeles Times* Woman of the Year 1970 – all the popular accolades that the U.S. of the 1970s could bestow.

But fame was just a means to an end. "There was this irresistible, almost patronising inclination to look at the women's team, that women had done this," she later observed, "so I – all of us – tried to twist the women's angle back to the real issue that the ocean was being noticed."[6]

Since then, over a long life of tireless work, Earle has raised public awareness about the plight of the world's oceans,

"THERE IS TIME – BUT NO TIME TO WASTE."

SYLVIA EARLE

as a speaker, film-maker, lecturer, government adviser, research scientist and above all an explorer with more than 7,000 hours spent underwater. In 2009, she set up Mission Blue, a Perpetual Planet Initiative partner, to inspire a worldwide network of marine protected areas called Hope Spots. Throughout her decades as a passionate and persuasive advocate for our oceans, Rolex has been her constant companion, whether meeting schoolchildren, heads of state or, of course, countless species of marine life. She has featured in numerous advertisements for the brand, most memorably coining the strapline: "My Rolex is even more at home on the floor of the Pacific than I am."

Over time, her understanding of the underwater world has deepened. "As a child, I accepted the widely held view that the ocean was limitless in terms of what people could take from it – or put into it, without causing harm. The world I knew seemed timeless. Now there is evidence that it has taken 4.5 billion years of complex geological, chemical and biological interactions to develop Earth into a planet suitable for humankind and it has taken humankind about four and a half decades to alter the temperature, chemistry and fabric of life in ways that threaten our continued existence."[7]

But as well as the perils of our times, she is also aware of its privileges. "It has been satisfying to be a part of the greatest era of discovery about the nature of life in the sea, to pioneer development and use of diving technologies that have enabled me to spend thousands of hours underwater seeing creatures there on their own terms – not dragged to the surface in nets and observed in laboratories."[8]

"Since I began exploring the ocean in the 1950s, I have been driven to do what I can do as a scientist, as a human being, to join with others to make a difference in terms of balancing human needs with human desires, and to treat the ocean and the rest of nature as if our lives depend on it – because they do. The Rolex on my wrist is a constant reminder of time on a practical, personal level – 'It's time to be awake!', 'It's time to go to work!', 'It's time for dinner!'. But it is also a symbol of time on a grand scale. There is time – but no time to waste – to use the unprecedented knowledge now available to secure an enduring place for ourselves, as a part of nature – not apart from it."[9]

Sylvia Earle, photographed by fellow Rolex Testimonee David Doubilet, observing sponge and coral growth on the pilings of a town pier. These man-made structures become artificial reefs that provide increased habitat for marine organisms.

SCUBAPRO

In 1960, Swiss oceanographer Jacques Piccard and U.S. Navy Lieutenant Don Walsh navigated the *Trieste* on a nine-hour plunge to the bottom of the Mariana Trench.

A NEW ERA OF DEEP-SEA EXPLORATION

At midnight, Andreas B. Rechnitzer started dropping charges of TNT over the side of the USS *Lewis* as it battled heavy Pacific swells. The remaining hours of darkness were punctuated by the boom of explosions. As each charge detonated just below the surface, sending a plume of spume into the air, Dr Rechnitzer started a stopwatch and waited expectantly as the seconds ticked by until his headphones crackled with the noise of the returning echo from the ocean bed. As the scientific director of the U.S. Navy's Project Nekton, he had been entrusted with the task of locating the deepest point of the ocean, and now, at six o'clock on the morning of 23 January 1960, 322 kilometres south of America's naval base on the island of Guam, he had found it.

Guam is the closest piece of land to the Mariana Trench, the truly awe-inspiring chasm that traces a graceful arc along 2,550 kilometres of ocean floor and contains the deepest waters on the planet. The echo from the last charge had taken 14 seconds to return, which, by his calculations, indicated that the seabed was an incredible 10,240 metres (33,600 feet) below his vessel's hull. The USS *Lewis* was right above the deepest point on the surface of the planet.

Within the next few hours, two men squeezed into a cramped steel sphere suspended under large tanks of ballast would descend far deeper beneath the waves than anyone had ever dived before.

For one of these men, a rangy six-foot-and-seven-inches-tall Swiss academic by the name of Jacques Piccard, this voyage to the deepest point under the world's oceans would complete a journey that had begun in, or rather under, the seas off Capri seven years earlier. As well as equipping Rebikoff, Delauze, Fructus and numerous other submariners, aqualungers, marine archaeologists and underwater explorers, not to mention the brand's work with the Royal Navy, Rolex had also been working with Swiss scientist, balloonist, bathyscaphist and explorer Auguste Piccard, as well as his son, Jacques, on a project to take a timepiece to even greater depths.

A friend of Einstein and the model for Hergé's Professor Calculus in the *Tintin* books, Auguste Piccard looked exactly as a distracted professor should: a long white coat, long white hair projecting like the upward-pointing eaves of a pagoda above his ears, and spectacles with an extra set of hinged lenses that could be flipped down when necessary. He may have been born in the days of Jules Verne, but by 1931 he had travelled by balloon to the edge of the stratosphere.

"THE COMPLETE PROGRAM OF A NEW SERIES OF UNDERWATER DIVES AT GREAT DEPTHS HAS JUST BEEN SET UP."

AUGUSTE PICCARD

Subsequently, he became fascinated by the conversion of his invention to the exploration of what became known as "inner space": the world under the waves. On 3 November 1948, an unmanned "15-ton sea balloon"[1] descended to a depth of 1,370 metres (4,500 feet) in the seas off Dakar near the Cape Verde islands: an experiment in which the then-little-known Jacques-Yves Cousteau had participated as part of the support team. With the successful descent and return of the vessel, Jacques Piccard observed that "a new era of deep-sea exploration had been opened."[2]

Inspired by the outcome of the dive, Auguste Piccard embarked on plans for something far more ambitious: a manned vessel capable of taking its occupants eight times deeper, to the very basement of the planet, the deepest point under the sea; a region at the time less well known than the trackless voids of space. As early as 15 December 1950, Auguste Piccard had written to Rolex about his plans for a bathyscaphe, but construction had run into delays so it was not until 25 February 1952 that Piccard issued a statement concerning the nature of the project.

"The complete program of a new series of underwater dives at great depths has just been set up.

"The construction of the bathyscaphe and a first series of experiments will be carried out within the framework of an Italian-Swiss organization under the auspices and with the help of the Italian Government, the Swiss Authorities in Rome, the Authorities and a museum in Trieste. The Italian Navy will participate in the research and will assist in all surface work. Most of the new bathyscaphe will be built in Italy and Trieste," after which the vessel would be named. "Part of the equipment and precision apparatus will be provided by Swiss firms. The technical management of this new construction has been entirely entrusted to me.

"Thanks to the help that was generously provided to me in Switzerland and to the contributions which have just been offered to me, in particular by the Government in Rome and the Authorities in Trieste, the necessary funds have been found, which allows the start of construction.

"I express my gratitude to all those who, by placing their trust in me, are now ensuring the realization of the project."[3]

Deep Sea Special, 1953, 37.5 mm, experimental watch featuring a domed crystal.

ROLEX

“WATCH PERFECTLY RESISTS DIVE 3,150 METRES.”

AUGUSTE PICCARD

Jeanneret was happy to hear from Piccard as it gave him the opportunity to revive a project, which, as suggested by correspondence in the Rolex Archives, could have first been mooted as far back as 1950.

“You will recall that we had planned to build, as an experiment, a waterproof watch containing a wristwatch-size movement, and that this watch would be attached to the outside of the bathyscaphe,” wrote Jeanneret to Piccard. “In all likelihood, this watch would take the shape of a sphere in order to better withstand the immense pressures. It would be fitted with a specially designed waterproof crystal, and if this were physically possible, the watch could be attached to the outside of a porthole and would be fitted with an extra-luminescent material dial in order to allow you, in theory at least, to read the time through the porthole while diving.

“We have of course no idea whether this is feasible in practice, and would be grateful if you could tell us how you imagine that an experiment of this kind could be undertaken with any chance of success. If it is impossible to attach the watch to the bathyscaphe itself, perhaps it could be attached to the cable that will support your machine?

“The watch movement would be fitted with an automatic winder, specially oiled to withstand very low temperatures. This movement would have a diameter of 32 mm, so that the complete watch would have a diameter of 40 to 42 mm. As we see things, it would be a sphere that would be smooth other than the means of attachment.

“We would be grateful if you could look into the matter, tell us what you think about it, and give us the benefit of your advice for the construction of this special watch, letting us know to what depths you feel you will be able to descend to, in order to enable us to make calculations regarding [water] resistance.”[4]

Jeanneret had to write to Piccard again to jog his memory, and it was not until 1 April that the scientist responded, with a letter written in his old-fashioned sloping hand.

“Since receiving your first letter I have examined the problem that you put to me from various angles. I am very happy that you got in touch with me regarding this question and it goes without saying that with you, I will do the best I can to achieve the result … We can very well place the watch in front of one of the portholes … It will thus be well protected, visible from the cabin, and exposed to external pressure. During the 1952 expedition in the Mediterranean, the temperature could vary between estimated values of +35 degrees to +5 degrees. In the Atlantic this would be plus +30–0 degrees centigrade.

"I have been meaning to write to you for some time to propose a construction for the special case that can resist water with complete security even under pressure of 100 to 200 metres. I think that one would be able to combine the two questions. The watch that has resisted pressures at 3,000 or 4,000 will make the best publicity for the current model of 100 metres."[5]

Jeanneret said that he would start work on a timepiece that, as soon as completed, would be "tested by the polytechnic in Zurich in order to give you the assurance that this watch will effectively withstand pressures of 3,000 or 4,000 metres without causing you any trouble.

"Consequently, we look forward to receiving the documentation that you want to prepare for us so that we can benefit from your vast knowledge of the subject."[6]

Reading these letters today is to be an eavesdropper on the first tentative discussions about the watch known to history as the Deep Sea Special. However, the use of the word "watch" should not suggest a portable personal timepiece suitable for quotidian use: in fact it made no concessions to real-world wearability. It was, to put it politely, a purposeful rather than an elegant object. The black dial, ringed by large luminous indices, carried the word "Rolex" in larger letters than usual in its lower half and a commensurately sized crown in the top half. With its protruding back and unusually deep case walls, the watch was far larger than usual. The winding crown, too, was oversized, but what really defined its appearance was the extraordinary watch glass. The domed crystal looked like a transparent table tennis ball that had been cut in half: the glass rose straight sided from the bezel for a few millimetres and then finished in a highly polished dome.

Following testing, a new, even more robust version was made with thicker crystal and a deeper, specially developed dodecagonal case back designed to be closed using tools capable of exerting greater torque than usual. Finally, on 30 September 1953, in the waters off Naples, the *Trieste*, with Auguste Piccard and his son Jacques aboard, dived to over three kilometres (1.86 miles) with a Deep Sea Special attached to the outside of the vessel. On returning to the surface, Auguste Piccard telegraphed Geneva: "Watch perfectly resists dive 3,150 metres best wishes – Piccard."[7]

Top: Swiss oceanographer Jacques Piccard in the conning tower of the *Trieste*, the deep-sea diving vessel designed by his father, Auguste Piccard. The bathyscaphe would become the first crewed submersible to reach the bottom of the Challenger Deep on 23 January, 1960. *Bottom:* The Deep Sea Special attached to the *Trieste*, circa 1957.

This page and following page:
Affixed to the outside of the bathyscaphe *Trieste* during the voyage into the Mariana Trench in 1960, the experimental Deep Sea Special successfully endured the colossal pressure that reigns at a depth of almost 11,000 metres (36,100 feet).

This summer dive in a newfangled type of submersible was to become a historic milestone for Rolex on its journey ever deeper beneath the waves. The equally outlandish-looking watch quickly became famous, and was the star attraction of the Rolex booth at the Basel Fair the following year, where the *Revue Française des Bijoutiers Horlogers* described it as "an article meriting particular attention".[8]

On 17 October 1956, Jacques Piccard took the *Trieste* and its externally mounted horological companion on another dive in the Tyrrhenian Sea, this time achieving a depth of 3,700 metres (12,140 feet). A grainy black-and-white photograph shows the watch attached to the hull of the vessel outside the access point to the hatch through which Piccard entered the steel observation sphere.

This feat came to the notice of the Director of the Museum of Watchmaking in La Chaux-de-Fonds, which requested the watch that had made this dive for its permanent collection. It was a request that Rolex was happy to grant, writing: "We are pleased to think that after the demanding test it has undergone, this chronometer will finish its career peacefully in the museum of watchmaking in La Chaux-de-Fonds."[9]

The museum was effusive in its thanks, and there was one line that would have pleased Wilsdorf as he read the letter: "This historic watch will certainly be appreciated by our visitors and, along with some other remarkable modern pieces, it will prove to each one that Swiss watchmaking remains worthy of its past."[10]

Nor was a watchmaking museum in the Swiss mountains alone in taking notice of these tremendous feats on the floor of the Mediterranean. News of the Piccards' achievements had reached the Pentagon. At the height of the Cold War, the potential of this deep-diving craft was of great interest to the U.S. Navy, which clearly saw a role for it in waters other than the Mediterranean. The Office of Naval Research purchased the *Trieste*, and on 20 December 1958, it made its first dive in the Pacific.

Between the end of 1958 and mid-May of the following year, the *Trieste* was out of the water undergoing an extensive rebuild, being fitted with new equipment and a new sphere capable of surviving dives of 10,973 metres (36,000 feet). On 8 January, the *Trieste* completed a five-and-a-half-hour dive into the lightless depths of the Mariana Trench during which the bathyscaphe reached the astonishing depth of 7,025 metres (23,048 feet).

The following day, a jubilant telegram arrived in Geneva.

*"VOTRE MONTRE TRES BIEN SUPPORTE PLONZEE [sic] 7300 METRES [sic] MEILLEURES SALUTATION = JACQUES PICCARD"** Transcription errors and slight inaccuracy regarding the final dive depth notwithstanding, this was extraordinary news.

In parallel with the U.S. Navy's enhancement of the *Trieste* bathyscaphe, Rolex had been making corresponding modifications to the Deep Sea Special to enable it to continue its underwater adventures at greater depths. The Deep Sea Special had already been tested to withstand pressures of 600 bar, equivalent to 6,000 metres (19,685 feet) – deeper than the deepest point in the Mediterranean. But for the extreme depths of the Pacific, a further iteration of this remarkable timepiece had been developed: the dodecagonal back remained the same, but the case had increased in size, the winding crown was wider and shorter, the bezel profile had been altered and, compared to the preceding version, the glass height had been lowered slightly, but was still 33 times higher than a normal crystal.[11]

In this strange, carbuncular timepiece was crystallized the sum of all the knowledge that Rolex had acquired in the field of waterproof watches since the appearance of the Oyster in the 1920s. At that time, the world had been stunned to learn that a Rolex had been worn on a cross-Channel swim; but by 1960, the world was no longer so easily impressed. The highest peak on Earth had been scaled, the sound barrier had been broken, the remotest corners of the planet had been mapped, man had flown to the edge of space, the skies were dominated by jet planes, new scientific frontiers were being crossed ... and Rolex had been the constant companion of mankind as these feats had been accomplished. There were no places on the planet's surface left to discover.

Yet, beneath the choppy seas of the Pacific Ocean there remained one last extreme location unvisited by man: the appropriately named Challenger Deep, a seven-mile-long, one-mile-wide gash in the floor of the already unusually deep Mariana Trench. The average ocean depth is around 3,700 metres (12,100 feet); Challenger Deep is three times as deep, and it was into this abyss that the *Trieste* and the Deep Sea Special were about to descend.

* "Your watch survived diving to 7,300 metres very well best wishes Jacques Piccard."

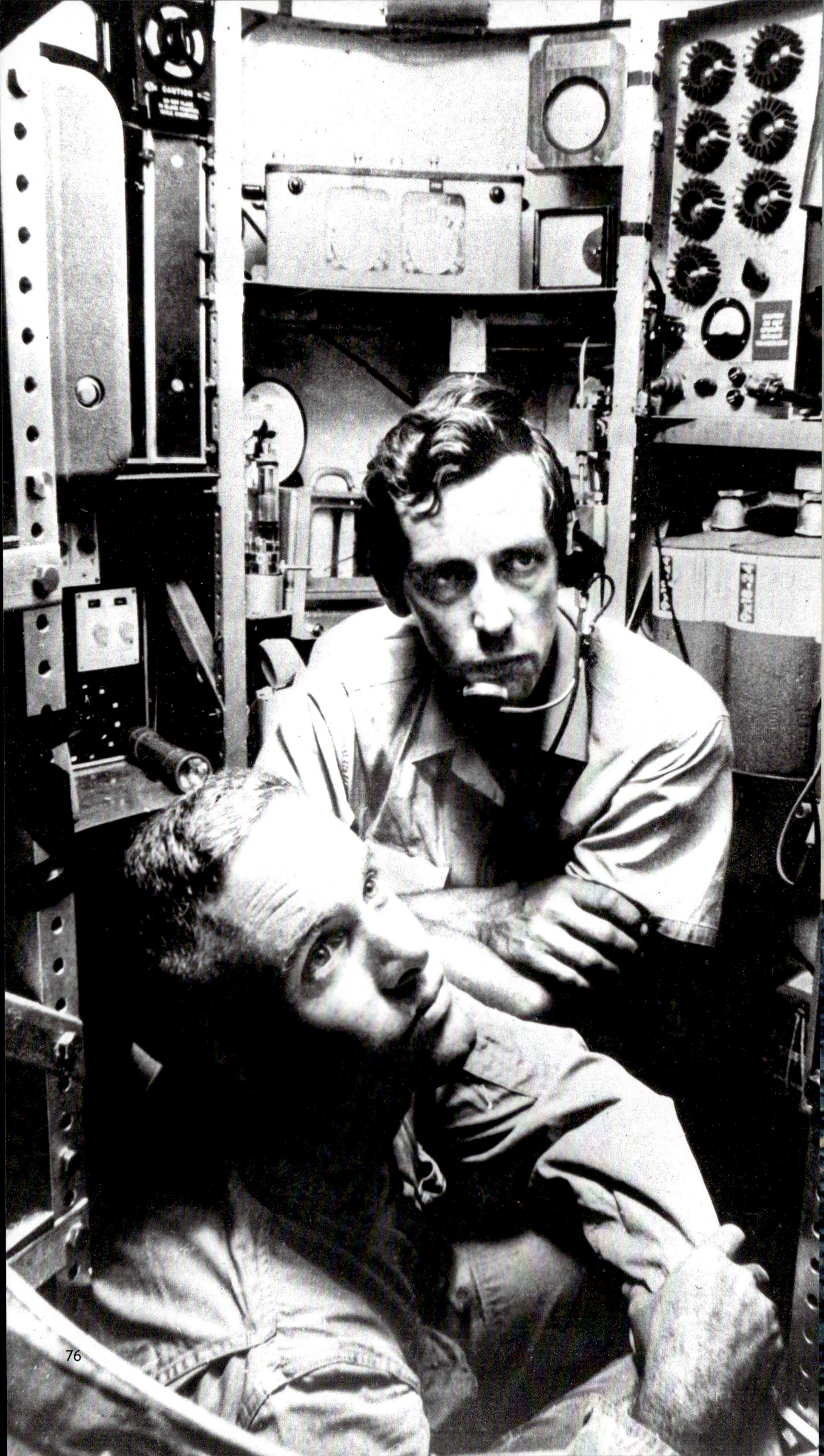

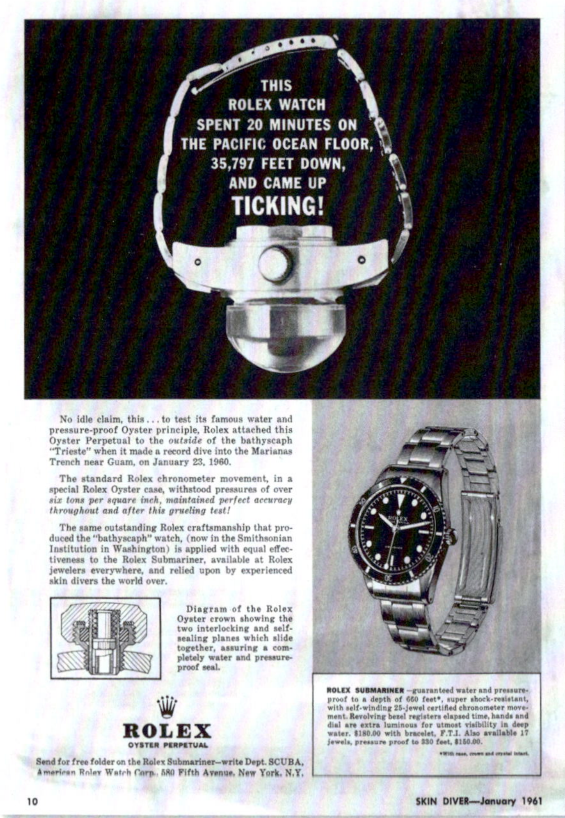

Rolex advertisement, 1961.

Previous pages: Jacques Piccard and Lieutenant Don Walsh inside the *Trieste* just before their nine-hour plunge to the Challenger Deep, the deepest known depression on the Earth's surface, January 1960.

By the time he boarded the wave-buffeted bathyscaphe from the support vessel, Piccard was drenched by the tempestuous seas, which had also wrought considerable damage upon the *Trieste*.

"The sight that met my eyes when we boarded the wallowing bathyscaphe was discouraging, to say the least. Broaching seas smothered her. The deck was a mess. Everything was awash. It was apparent at once that the tow from Guam had taken a terrific toll.

"A hasty inspection revealed that the surface telephone (used for topside communication after we were sealed in the sphere) had been carried away. The tachometer (the instrument that measured our diving rate) was badly damaged and inoperative. The vertical current meter was dangling by a few wires.

"I was worried."[12]

But before making a final decision as to whether to proceed or postpone, he wanted to see the state of the equipment within. He hoisted himself into the conning tower, his sodden shoes squelching on the swaying 18-foot-long ladder to the claustrophobic metal ball, its walls lined with instruments. He powered them up and all were in order. He consulted the Rolex on his wrist: it was now 8 a.m.; leave it much later and he risked not returning to the surface until after dark. He decided to make the dive, and for the next few hours his world would shrink to the pocket of air in the cramped steel ball "38 inches square and only five feet, eight inches high", which he would share with his co-pilot, Naval Lieutenant Don Walsh, as they made their journey to the bottom of the sea.[13]

The hatch, resembling the entrance to a stout bank vault, was lowered into place. Its bolts were tightened until the grease squeezed out of the cracks. Within minutes, two tonnes of seawater had flooded the ballast tanks.

The *Trieste* sank beneath the boisterous waves.

Descent and ascent were controlled by seawater and iron shot used as ballast, along with gasoline for buoyancy as it is lighter and less dense than water. At 91.5 metres (300 feet), some of the gasoline was jettisoned to pass the thermocline where the water temperature drops sharply, making the water denser. At 183 metres (600 feet), what little light there was began to fade; by 305 metres (1,000 feet) it was gone and the *Trieste* was dropping through the night-black water at the rate of about 1.2 metres (4 feet) per second.

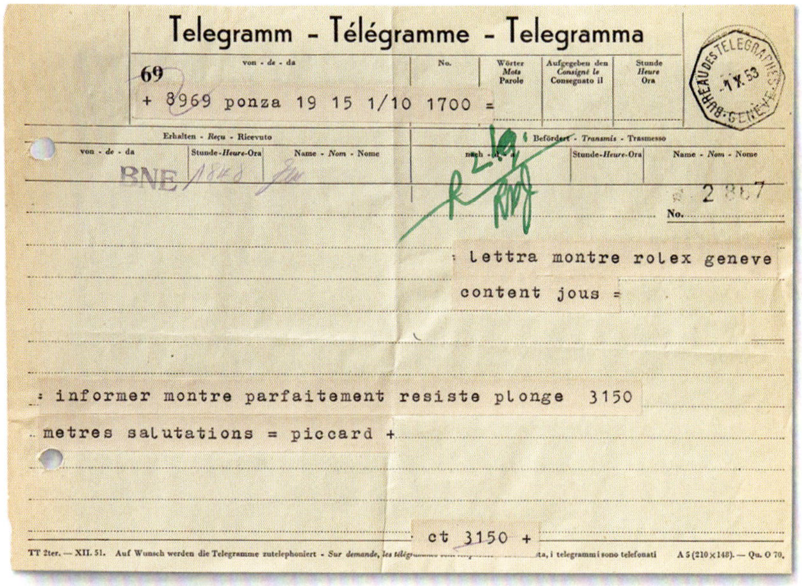

Telegramm - Télégramme - Telegramma

69

+ 8969 ponza 19 15 1/10 1700 =

BNE

2 867

lettra montre rolex geneve

content jous =

: informer montre parfaitement resiste plonge 3150

metres salutations = piccard +

ct 3150 +

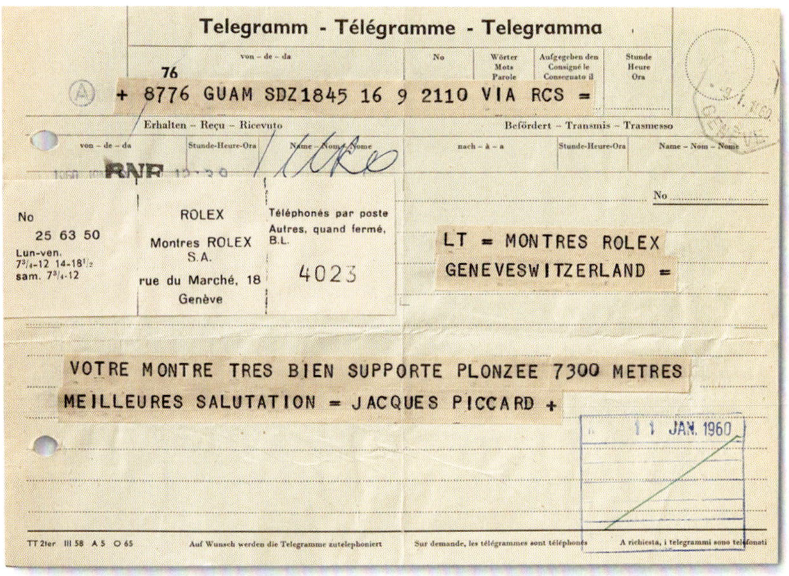

Telegramm - Télégramme - Telegramma

76

+ 8776 GUAM SDZ1845 16 9 2110 VIA RCS =

BNE

No 25 63 50

ROLEX Montres ROLEX S.A. rue du Marché, 18 Genève

Téléphonés par poste

4023

LT = MONTRES ROLEX

GENEVESWITZERLAND =

VOTRE MONTRE TRES BIEN SUPPORTE PLONZEE 7300 METRES

MEILLEURES SALUTATION = JACQUES PICCARD +

11 JAN. 1960

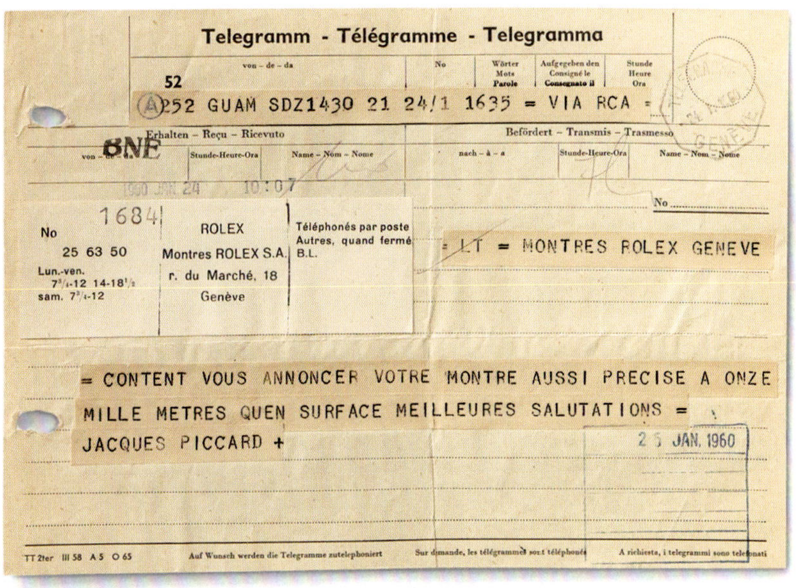

Telegramm - Télégramme - Telegramma

52

252 GUAM SDZ1430 21 24/1 1635 = VIA RCA =

BNE

1684

No 25 63 50

ROLEX Montres ROLEX S.A. r. du Marché, 18 Genève

Téléphonés par poste

LT = MONTRES ROLEX GENEVE

= CONTENT VOUS ANNONCER VOTRE MONTRE AUSSI PRECISE A ONZE

MILLE METRES QUEN SURFACE MEILLEURES SALUTATIONS =

JACQUES PICCARD +

25 JAN. 1960

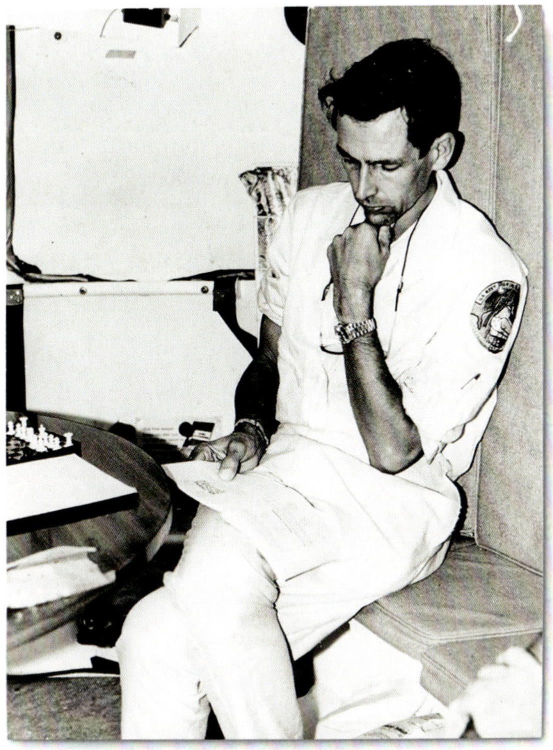

IN THIS STRANGE, CARBUNCULAR TIMEPIECE WAS CRYSTALLIZED THE SUM OF ALL THE KNOWLEDGE THAT ROLEX HAD ACQUIRED IN THE FIELD OF WATERPROOF WATCHES.

Telegrams from Jacques Piccard *(above)* to Rolex heralding their depths.

THE JOURNEY TO THE SURFACE TOOK THREE HOURS AND 27 MINUTES; TO THE MEN IN THE *TRIESTE* – COLD, TIRED AND EMOTIONALLY DRAINED – IT SEEMED MUCH LONGER.

External pressure started to intensify, and at 3,000 metres (10,000 feet), a small leak started, but Walsh and Piccard were familiar with its nagging "drip, drip, drip" from previous dives. They predicted it would stop at around 4,500 metres (15,000 feet), which it did. They had not even reached the halfway point of their journey to the deepest point on the planet. At around the same point, they lost voice contact with the support vessels; from then on, crackling was all they heard through their headphones. But for their puny lights, the hiss of the air system and the hum of the electronic equipment, they were alone, plummeting through a world of silence and darkness. At around 9,144 metres (30,000 feet), that silence was shattered. "We heard and felt a powerful, muffled crack. The sphere rocked as though we were on land and going through a mild earthquake," Walsh later recalled.

The next seconds were agonizing. "We waited anxiously for what might happen next. Nothing did. We flipped off the instruments and the underwater telephone so that we could hear better. Still nothing happened. We switched the instruments back on and studied the dials that would tell us if something critical had occurred. No. We had our equilibrium and were descending exactly as before."[14]

Should they return to the surface? They had no way of telling how damaged their vessel was and whether it could complete the rest of the journey. But the ocean floor was tantalizingly close, by their calculations just 1,067 metres (3,500 feet) away.

They decided to continue, albeit at a slower speed. When they believed that they were 183 metres (600 feet) from the bottom, they turned on the sensitive fathometer to gauge the final distance from the seabed so they could fine tune their "landing".

It was then that things began to get really strange.

It registered nothing.

Walsh was incredulous. "We continued to slide on down. It also showed nothing 100 feet later – or 100 feet below that."[15] Piccard attempted to lighten the mood. "At 36,000 feet, Jacques asked me wryly whether I thought we could have missed the floor somehow. I said I thought the chances were against it.

"We checked our speed to half a foot a second and continued. At that rate, time and distance pass very slowly, and I think for the first time in the dive, both of us had the feeling of awe that comes from exploring the totally unknown.

"I did not take my eyes off the fathometer and Jacques never stopped watching out of the tiny porthole with its weak probe of light. No bottom was in sight at 36,600 feet, or at 37,200. But at last at 37,500 feet the fathometer traced the beginnings of the bottom. Soon Jacques could see a difference in the effect of our light in the water, as the rays reflected off the bottom. As we approached the floor, I called the fathometer readings to Jacques in fathoms: 'Thirty ... twenty ... ten ...' At eight, he called that he could see the grey-white bottom."[16]

Although no filmed record of this adventure exists, hearing Walsh count down the final moments of their descent, the mind's eye recalls the now-familiar grainy footage shot from the *Apollo* landing craft as it came to rest on the lunar surface. But the lunar landing was nine years in the future, and at that moment, Piccard and Walsh were at the very limits of human knowledge, about to set their craft down on land that had never before been seen by human eyes, let alone visited by a human being.

With their instruments reading a stunning, if inaccurate, 11,520 metres (37,800 feet)*, the *Trieste* came to rest in a cloud of ash-coloured silt, "indifferent to the nearly 200,000 tons of pressure clamped to her metal sphere."[17]

To celebrate, Piccard unfurled a Swiss flag and Walsh the Stars and Stripes, pictures were taken, hands were shaken. And then Walsh decided to take a look through the aft porthole and was concerned to see "a series of large cracks, stretching from one side to the other."[18]

It was time to leave.

The journey to the surface took three hours and 27 minutes; to the men in the *Trieste* – cold, tired and emotionally drained – it seemed much longer. Their sense of bathos gave way to euphoria on the surface as, teeth chattering, they scrambled up the ladder into the sunlight and salt-scented air.

"We seemed to have arrived on the surface in the middle of a carnival. Two navy jet photo planes blasted by us a few feet above the conning tower. An Air Force search and rescue C-54 made several low passes at us. Off to the west, the *Lewis* was bearing down on us, and right behind her came the *Wandank*, breaking all speed records for tugs. She looked like a 30-knot destroyer heading for a fire. She looked great. So did everything else."[19]

The men were no longer scientific researchers – they were conquering heroes. President Eisenhower presented them with awards at the White House, and when John F. Kennedy was elected president, the *Trieste* was part of the inaugural parade, hauled through the streets of Washington mounted on a trailer emblazoned with the proud words "Navy Conquers Inner Space" together with the adjusted distance "35,800 Feet Down". The Rolex that had been attached to the outside was presented to the Smithsonian Institution. Such was the demand for examples of the watch that a number of demonstration models were made for exhibition at retailers across the globe.

This remarkable feat made news around the world, especially in Geneva, where at seven minutes past 10 o'clock on the morning of 24 January, the teleprinter in Geneva's telegraph office clattered into life with the news from Guam.

*Content vous annoncer votre montre aussi précise à onze mille mètres qu'en surface. Meilleures salutations. Jacques Piccard***[20]

Surviving prolonged immersion in waters barely above freezing point, resisting pressures of up to 1.15 tonnes per square centimetre, a Rolex had travelled to and returned from the deepest depths of the ocean, functioning as perfectly as if in the workshops in Switzerland where it had been made.

* The reading was inaccurate and later judged to be 10,916 metres (35,814 feet).
** "Glad to report your watch as accurate at 11,000 meters as on land. Best wishes. Jacques Piccard."

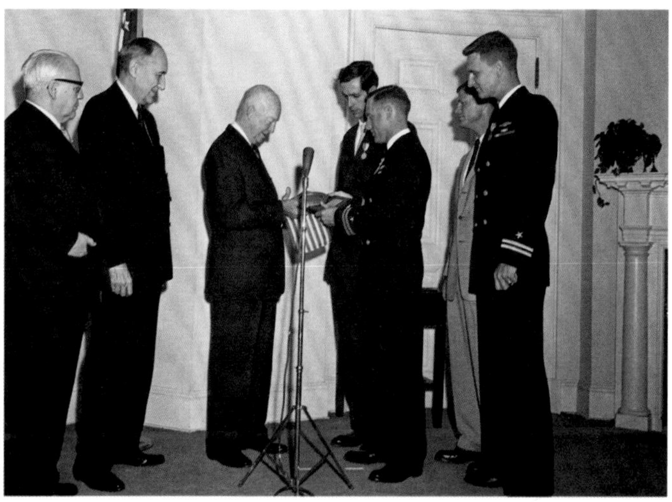

President Eisenhower awards Jacques Piccard and Lieutenant Don Walsh with medals.

Henri Delauze was a pioneer in the science and practice of saturation diving. He devised technical solutions to difficulties encountered at extreme depths.

DAY OF THE AQUANAUTS

Although it had remained on the ocean floor for a scant 20 minutes, Dive No. 65 by the *Trieste* to the bottom of the Challenger Deep marked a transformational point in humanity's relationship with the undersea environment. A bona fide scientific and military expedition mounted by the world's acknowledged economic, scientific and military superpower, it had claimed the deepest point on Earth for the U.S. and the free world. After presenting Piccard with his award, Eisenhower wrote to him to express the nation's gratitude for his "outstanding contribution to the United States and science in the field of oceanographic research.

"As a citizen of Switzerland, a country admired by all the free world for its love of freedom and independence, you have the gratitude of all of the people of the United States for helping to further open the doors of this important scientific field."[1]

The world seemed poised on the threshold of a new era of discovery. By the beginning of the 1960s, the nature of underwater exploration had changed from thrill-seeking sub-aqua safari, or archaeological treasure hunt, to something more profound, affecting national security and the very future of humankind.

"It is safe to predict that within a few years, the rush into outer space will be matched by an international invasion of inner space which is even now vaguely shaping up. The some 350 million cubic miles of water space cannot remain neglected. It is a large living space in comparison to all other realms of life on Earth combined,"[2] wrote oceanographer Robert S. Dietz in 1961.

Four years earlier, he had been the first American to take a deep dive on the *Trieste* off Capri with Piccard, and it had been an experience of Damascene intensity.

In 1961, Henri Delauze had opened a deep-sea diving business in the Old Port in Marseille called Compagnie Maritime d'Expertises, which would soon be better known by the acronym Comex.

Delauze was part pioneer and part businessman: "The ability to send divers to work at deeper depths might prove to be good for business," observed one commentator, "but for Delauze, depth barriers were there to be broken, and he relished the prospect of having his company break them first."[3] In order to help him take humans ever deeper beneath the waves, he assembled a team of experts: one early recruit to Comex was Jeanneret's friend, Dr Xavier Fructus, who had also worked for Cousteau but was lured away by Delauze to carry out medical research at Comex into the physiological effects of deep diving.

The timing for Delauze's new business could not have been more propitious. Events occurring simultaneously on the other side of the Atlantic would propel Comex – and Rolex – to the forefront of the next phase of underwater exploration.

In 1961, the oil industry underwent a truly Archimedean change. Blue Water Rig No. 1 was drilling in the Gulf of Mexico on behalf of Shell. But, when the time came to move it, "the pontoons were not sufficiently buoyant to support the weight of the rig and its equipment, it was

JUST AS ROLEX CREATED THE GMT-MASTER USED BY PILOTS OF JET AIRCRAFT, SO IT DEVELOPED A WATCH FOR THE NEEDS OF SATURATION DIVERS.

towed between locations at a draft midway between the top of the pontoons and the underside of the deck. It was observed that the motions at this draft were very small and Blue Water Drilling and Shell jointly decided that the rig could be operated in the floating mode."[4]

Almost by accident, offshore oil exploration had crossed a technical Rubicon. It was no longer necessary for an oil platform to stand on the seabed. Thus liberated, oil companies were able to drill in much deeper waters using floating oil platforms. The first such purpose-built rig, Ocean Driller, entered service two years later and changed the industry. Noted for its stability, Ocean Driller was a V-shaped structure. It may look unfamiliar now, but its principles have informed oil rig design ever since.

Using semi-submersible technology, it became possible to extract oil from well heads up to 2,000 metres (6,562 feet) under water. As the oil industry expanded into such areas as the North Sea, it brought with it transformative riches that changed once-remote areas such as Scotland's east coast and Norway's Stavanger peninsula into new Klondikes. Coastal cities, including the once-busy fishing port of Aberdeen and Stavanger, the quondam fish-canning capital of Scandinavia, became boom towns. The effect of this new technology on the oil industry can be likened to that of jet aircraft on aviation and, just as Rolex created the GMT-Master used by the pilots of jet aircraft, so it developed a watch for the needs of the action heroes of this new and lucrative but hazardous industry: saturation divers.

"Saturation (sat) diving is when the inert gas breathed by a diver dissolves into the body's tissues and reaches equilibrium with the ambient pressure at the diver's depth (i.e. no more gas can be absorbed by the tissues – they're fully saturated). This is Henry's law, named for British chemist William Henry,"[5] explains Sherri Ferguson, diver, scientist and Director of the Environmental Medicine & Physiology Unit at Simon Fraser University.

"Recreational divers limit their time at depth to avoid becoming oversaturated, so they can make a direct ascent to the surface without any mandatory stops. Upon surfacing, they're supersaturated, meaning the amount of inert gas in the body is greater than the amount found in the surrounding atmosphere. Nature likes balance. The inert gas comes out of the tissues and into the blood,

Henri Delauze during the exploration operation Sud Caveaux I, south of Pointe Caveaux off Marseille, 1996.

U.S. NAVY
SEALAB
SEALAB II

which circulates to the lungs, where the diver exhales it. A technical diver who makes a dive beyond recreational limits must make decompression stops to avoid overburdening the body's ability to off-gas. For saturation divers, these stops can take a week."[6]

The breakthrough in understanding the behaviour of the human body at extreme depths was made by Captain George Bond, director of the Medical Research Laboratory at the New London Connecticut submarine base: a large, hearty figure of a man who, when he did not have an underwater breathing apparatus in his mouth, was to be seen with a pipe clamped between his teeth. Born in 1915, Bond was, not unlike Delauze and Cousteau, a maverick, or as he put it with colourful figurative flourish, "the oddball who loses sight of the ultimate military goal and shows a disturbing tendency to pick daisies during the forced march."[7]

Before joining the navy, he had been a country doctor living in a log cabin with his young family in the evocatively named small town of Bat Cave (population 170, give or take), located midway down the Hickory Nut Gorge in the Blue Ridge Mountains of North Carolina, a location so bucolically American that one can almost hear the wheezing accordions, the heehaw of fiddles, plinking banjos and tapping feet of bluegrass music. In practising mid-20th-century medicine, he encountered some decidedly 19th-century attitudes. On one occasion, an old woman aimed her shotgun at him when he prepared to inject a relative with penicillin. Fortunately, he had taken the precaution of making his patient visits with a .38 mm pistol in his doctor's bag and, penicillin jab in one hand, with his other he pulled out his handgun and levelled it at the old woman. He also doubled as a lay preacher, built a hospital, and, shotgun-toting grandma notwithstanding, he clearly won his patients over, becoming something of a national celebrity, written about in the multimillion-selling circulation *Reader's Digest* and appearing on the television programme *This Is Your Life*.

In 1953, the Korean War nearing its end, he was called up as a paratrooper doctor, but with the armistice in July that year, he transferred to the submarine service on the grounds that the bonus "hazard" payment was identical to what he would have earned as a paratrooper. It is on such apparently trivial accidents that history pivots. "Five days later, I found myself at the U.S. Naval School of Deep Sea Divers, not exactly on my planned agenda. I was not only surprised to be there, but also delighted to find that I loved to dive. This fascination with diving has stayed with me all my life,"[8] he later recalled in his journals.

His newfound love of the underwater world stimulated his medical curiosity, and it was through talking to Jacques-Yves Cousteau that he began to form his theory of deep-sea diving. "From that conversation, Captain Bond developed the process whereby the body's tissues are saturated with helium so divers could withstand the prolonged pressure of deep ocean dives,"[9] observed his 1983 obituary in *The New York Times*. He concluded that "once saturated, the diver's decompression can be based on depth rather than duration of the dive. Thus, a diver saturated to 300 feet would need the same decompression time (about two-and-a-half days) whether the stay was one day or one month."[10]

In order to prove that humans could live and work on the seabed for extended periods of time, Bond assembled a team of elite navy divers. First, they lived in a land-based pressurized tank to simulate underwater pressure. Once the theory had been tested to Bond's satisfaction, the experiment was moved underwater to a series of increasingly elaborate ocean floor habitats: Sealabs I, II and III, where a carefully selected group of men would live and work as part of an extended experiment in underwater existence.

Set apart from others and bound together by the perils they faced, this was a special group of men, an anointed few who displayed the physical aptitude, experience, courage, daring and skill required to venture beneath the waves and then stay there for extended periods. Like their near-homonyms the space-travelling astronauts, these so-called "aquanauts" walked into history with a special swagger. Part explorers and part laboratory animals, their heads shaved, electrodes glued to their scalps and on their bodies, aquanauts were subjected to extensive medical testing before descending to their underwater home, where they would spend weeks at a time, performing experiments and taking readings from each other as data and information about how the human body adapted to life underwater. It was the aquanauts' bravery, and Bond's desire to pick metaphorical daisies, that made the offshore oil industry viable.

Sealab II, an experimental underwater research laboratory developed by the U.S. Navy and placed off the coast of La Jolla, California, 1965.

Rolex advertisement featuring the Comex expedition Hydra VIII, 1988.

As oil wells moved further offshore into deeper waters out of sight of land, the requirement for skilled divers capable of carrying out technical tasks increased. Given the lengthy decompression at a rate of approximately one day per 30 metres (100 feet) of seawater, saturation diving made much more sense if, once at working depth under great pressure, their tissues saturated with a mixture of helium and oxygen, the divers could remain there. But for a commercial oil drilling company, the expense, risk and difficulty of constructing, maintaining and moving an underwater residential complex seemed overwhelming.

It was Comex that devised the solution.

Instead of divers living on the seabed, Delauze brought the seabed to the surface. From the mid-1960s, large metal tanks with airlocks, their sides perforated by small, thickly glazed porthole-like windows, began to appear on the new generation of semi-submersible rigs. Comex was supplying oil-drilling platforms with surface-level hyperbaric chambers that would serve as pressurized living quarters for divers, who would then "commute" via diving bell to their working depths, allowing a month-long tour of duty to be completed with just one multi-day decompression session.

Nor were these hyperbaric chambers stationed solely on oil rigs. At Comex headquarters, dives to ever deeper and deeper depths were simulated in advanced hyperbaric chambers, allowing Delauze, Fructus and Rolex to devise technical and procedural solutions to any difficulties encountered at extreme depths. It is fair to say that this was a journey into terra – or, perhaps it is more precise to say aqua – incognita that required the same sort of preparation as the space programme.

First there was the crucial question of what the aquanauts would breathe metres below the waves. The use of heliox (a mixture of helium and oxygen) for saturation diving was soon well established. Before their application to diving, helium mixtures had been used medically: easier to breathe and absorb, they required less effort to ventilate the lungs because helium is less dense than air and reduces resistance to flow in the airways.

Helium molecules are small; so small, in fact, that human airways are not the only things they enter easily. It was the diminutive size of those molecules that at 5.30 p.m. on 6 November 1967 gave André Zibach cause to trouble the Swiss patent office to protect another Rolex invention with an application for a patent (No. 492246), subsequently granted in the summer of 1970, for "a waterproof watch comprising a valve mounted in a passage in the case to communicate the interior space thereof with the exterior."[11]

The patent application acknowledged that watch cases with valves already existed.

"However, in existing constructions, the purpose of the valve is to permit either injection of a gas into the case, in order to make internal pressure higher than the ambient pressure, preventing penetration of water, steam or dust inside the case, or on the contrary, to create a vacuum inside the case, in order to reduce the effect on the movement of the air trapped in the case when it is closed, or with a view to improving the water-tightness of the case by the ambient pressure contributing to pushing its various constituent elements against each other.

"However, these constructions have never been used to eliminate the effects, on the watch case, of a long stay in a high pressure gaseous environment.

"Indeed, we have noted that, whatever the sealing qualities of the watch cases produced up to now, when they are subjected for relatively long periods, of the order of several hundred hours, to relatively strong pressures, of the order of several tens of atmospheres, and especially when the ambient atmosphere is made of a gas with molecules of small dimension, as it is the case of helium for example, which is used frequently in diving bells, the pressure inside the case eventually equilibrates with the ambient pressure.

"Such conditions of employment are not hypothetical but exist in reality, especially when the watch is used at great depths underwater, during work being carried out using a bell.

"During the return of the watch to a normal pressure atmosphere, and despite the decompression stops which are necessary for the occupants of the bell, there is an excess of pressure in the watch case, which can cause it to 'explode', that is to say, in particular, eject its glass."[12]

Arid and unexciting though the language of this official document may be, it can be considered as the birth certificate of one of the most remarkable watches of the latter half of the 20th century; a timepiece that would add further lustre to the name Rolex and confirm its hegemony in the production of what were effectively over-engineered, effectively indestructible watches capable of surviving conditions that would be lethal to the wearer.

During the summer of 1966, development had begun on a high-performance version of the Submariner: a watch that would accompany diving into a new era, and divers to new depths. A confidential report on a product meeting that took place in late September 1966 sheds light on the very conception of what, for almost 50 years, was to be the most extreme expression of Rolex's Professional watches. The memo lists new and updated models including the Submariner Seadweller 1000; written by hand is the reference number 1665. Fitted with the calibre 1575, the dial would carry the wording "Rolex Submariner Sea-Dweller 1000 – Superlative Chronometer – Officially Certified."[13]

"New!" proclaims the text under the heading of case and bracelet, and it was noted that, "if necessary, there is the possibility of removing the Cyclops magnifier", and that the watch was to be made in steel and possibly in gold.[14] As it transpired, it would be many decades before gold would appear on the Sea-Dweller, and even then, it would be combined with steel in a Rolesor version (Ref. 126603).

Work progressed swiftly, and on 9 May 1967, the executive committee overseeing design met to discuss the watch and decided that a first series of at least 10 Sea-Dwellers should be progressed "as swiftly as possible". Tests were to be made regarding a bidirectional ratcheting bezel, but for the 10 pieces that were urgently required, the bezel system already employed on the Submariner 200 m could be used. After a further meeting on 23 May, it was decided to "abandon the idea of a Cyclops glass for this model. In fact, this glass will be replaced by a glass capable of withstanding 50 atmospheres of pressure."[15]

A single atmosphere corresponds to approximately 10 metres (33 feet) of depth underwater. At 50 atm, or approximately 500 metres (1,640 feet), the water pressure is approximately 50 kg/cm^2 or 725 psi. These figures may seem somewhat abstract, so it should be remembered that, at that time, only a dozen scuba divers had ever dived to a depth of 240 metres (787 feet). Now, Rolex was designing a production timepiece capable of submersion to twice that depth, to waters where barely one per cent of surface light penetrates and the temperature is usually just a few degrees Celsius. The Sea-Dweller that went on sale was capable of operating at a depth of 610 metres (2,000 feet).

FOR ALMOST 50 YEARS,
THE SEA-DWELLER WAS
TO BE THE MOST EXTREME
EXPRESSION OF ROLEX'S
PROFESSIONAL WATCHES.

Sea-Dweller, 2019, Ref. 126603,
43 mm, yellow Rolesor version, waterproof
to 1,220 metres (4,000 feet).

1922

UNDER THE SURFACE

SUBMARINE

In 1922, Rolex launched the Submarine – a watch attached on a hinge inside a second, outer case, whose bezel and crystal were screwed down to make the outer case watertight. Accessing the crown – to wind the watch or set the time – required opening the outer case. The Submarine marked the first step in Hans Wilsdorf's efforts to create a completely sealed watch case that was convenient to use.

1926

UNDER THE SURFACE

OYSTER

Launched in 1926, the Oyster was the world's first waterproof wristwatch thanks to its Oyster case.

1953

100 M/330 FT

SUBMARINER

The Oyster Perpetual Submariner (38 mm), the first divers' wristwatch guaranteed waterproof to a depth of 100 metres (330 feet).

1953

3,150 M/10,335 FT

DEEP SEA SPECIAL

In 1953, in the waters off Naples, the *Trieste* with Auguste Piccard and his son Jacques aboard, dived to a depth of over three kilometres with this version of the Deep Sea Special attached to the outside of the vessel.

1954

200 M/660 FT

SUBMARINER

In 1954, Rolex went on to make further technical advances that rendered the Submariner waterproof to a depth of 200 metres (660 feet).

1960

10,916 M/35,814 FT

DEEP SEA SPECIAL

On 23 January, the bathyscaphe *Trieste*, piloted by Swiss oceanographer Jacques Piccard and U.S. Navy Lieutenant Don Walsh, descended 10,916 metres (35,814 feet) into the Mariana Trench. An experimental Rolex watch, the Deep Sea Special, was attached to the exterior. Upon returning to the surface, the watch was found to be in perfect working order.

EVER SINCE THE INVENTION OF THE OYSTER CASE, ROLEX HAS CONTINUED TO PUSH THE BOUNDARIES OF RESEARCH IN WATERPROOFNESS AS PART OF ITS RELENTLESS PURSUIT OF PERFECTION.

1967

610 M/2,000 FT

SEA-DWELLER

Launch of the Oyster Perpetual Sea-Dweller 2000 (40 mm), a watch designed for the pioneers of professional deep-sea diving, guaranteed waterproof to a depth of 610 metres (2,000 feet).

1978

1,220 M/4,000 FT

SEA-DWELLER

Launch of the Oyster Perpetual Sea-Dweller 4000 (40 mm), guaranteed waterproof to a depth of 1,220 metres (4,000 feet), with a sapphire crystal, Triplock winding crown and helium escape valve.

2008

3,900 M/12,800 FT

ROLEX DEEPSEA

Oyster Perpetual Rolex Deepsea (44 mm), guaranteed waterproof to a depth of 3,900 metres (12,800 feet). This ultra-resistant divers' watch benefits from three Rolex innovations: the Ringlock system, an innovative case architecture; the Chromalight display, introducing an exclusive luminescent material emitting a long-lasting blue glow; and the Rolex Glidelock bracelet extension system.

2012

12,000 M/39,370 FT

ROLEX DEEPSEA CHALLENGE

On 26 March, attached to the manipulator arm of James Cameron's submersible, the experimental Deepsea Challenge watch accompanied the film-maker and explorer on his solo dive into the Mariana Trench, reaching a depth of 10,908 metres (35,787 feet).

2022

11,000 M/36,090 FT

DEEPSEA CHALLENGE

Oyster Perpetual Deepsea Challenge, crafted from RLX titanium; 50 mm in diameter, waterproof to 11,000 metres (36,090 feet), with Ringlock system and helium escape valve. It is the first watch in RLX titanium to appear in the Rolex catalogue.

Astronaut-turned-aquanaut Scott Carpenter atop Sealab II, in which he lived at a depth of 61 metres (200 feet) for 30 days in August 1965.

THE NEXT FRONTIER

René-Paul Jeanneret was soon to depart for the U.S., where he had been invited to the 50th anniversary celebrations of the Langley Research Center, a long-established experimental technology facility in Virginia that was now part of NASA.

"I had a most fascinating time in the USA," wrote Jeanneret on his return in October 1967. "Besides Washington, I visited the Langley Research Center on the occasion of the 50th anniversary of NASA, took a grand tour of Cape Kennedy and received a warm welcome, so typical of traditional American hospitality, from some astronaut friends at the Houston Manned Spacecraft Center." Although not the official-issue timepiece of the Apollo Space programme, Rolex watches were widely worn by top airmen and astronauts. "I came back with a wealth of new ideas which I hope will prove useful to Rolex in the not too distant future, and needless to say, I was most impressed with all I saw."[1]

Today it would be called a networking opportunity, and as a networker par excellence, Jeanneret returned to Geneva having made valuable connections, among them astronaut Scott Carpenter, whom he had sounded out about testing a new watch.

"I am quite sure there will be some way to conduct an evaluation of your "Sea Dweller" [sic] on a non-interference basis, but to an extent that will provide us both valuable information,"[2] wrote Carpenter to Jeanneret in November.

The second American astronaut to orbit the Earth in *Mercury-Atlas 7* on 24 May 1962, Scott Carpenter was one of the many remarkable individuals who came to prominence during the frenzied years of the Space Race. However, by the time he met Jeanneret in 1967, he was more interested in the world beneath the waves than the voids of outer space. He had heard Cousteau speak at the Massachusetts Institute of Technology and had been so inspired that he contacted Captain Bond, took a leave of absence from NASA and joined the Sealab Project.

Bond recognized that he had in Carpenter a high-profile ambassador brimming with a convert's zeal, and in August 1965, the astronaut-turned-aquanaut was made mission commander of Sealab II, living at a depth of 61 metres (200 feet) for 30 days. The effect was profound. It was, said Carpenter, "the most richly satisfying experience of my life".[3] Shortly before leaving Sealab, he received a call from President Lyndon Johnson and, returning to the surface, he expressed the opinion that it was the patriotic duty of all Americans to support the Sealab project. "An important national frontier lies just under the surface of the sea, and we must get down to it and stay there."[4]

Sealab could have found no better advocate than the clean-cut Carpenter – and exactly the same could be said of the Sea-Dweller. The Rolex Archives contain a photograph showing a sharply suited Carpenter, his eyes covered with aviator sunglasses, looking like he was auditioning for a role in the *Men in Black* film franchise, thrusting his Rolexed wrist to camera. He and Jeanneret established an excellent rapport, and by 1968, the Sea-Dweller had been cleared for use, testing and evaluation on Sealab III.

It was a spectacular opportunity, and a watch was dispatched as soon as possible. "The Sea-Dweller arrived in time to be worn by one of our divers who saturated at 825 feet and made an excursion dive to 1,025 feet," wrote Carpenter excitedly on 27 February. "As far as I know it performed well throughout. The divers are not yet back to the surface and more information will be available shortly."[5]

"THE ROLEX SUBMARINER AND ROLEX SEA-DWELLER HAVE A CACHET THAT IS UNMATCHED BY ANY OTHER WATCH. THE ROLEX BRAND IS THE MOST HIGHLY RESPECTED IN MY BOOK. THE ROLEX SUBMARINER WAS THE PREFERRED DIVING WATCH FOR U.S. NAVY SEALAB DIVERS."

SCOTT CARPENTER

NASA astronaut Scott Carpenter, the prime pilot for the *Mercury-Atlas 7* flight at Cape Canaveral, Florida, conquered both outer and inner space. As an aquanaut and member of the U.S. Navy Sealab programme, Carpenter was part of the aquatic pioneering team that tested the Sea-Dweller model.

Sealab III was to be a much more ambitious project than its predecessors. Sealabs I and II had been located at depths of 59 metres (192 feet) and 63 metres (205 feet) respectively, but for the third mission, a refurbished Sealab II would be sent 186 metres (610 feet) underwater off San Clemente Island. Operations finally got underway in February 1969, but, just a few days in, an aquanaut died during an attempt to repair a dangerous helium leak and the project was terminated early.

The mission's failure came amid the spiralling costs of both the Vietnam War and the Apollo space programme, and Sealab was shelved. As the project was being wound up, the U.S. Navy's Head of Ocean Engineering returned the Sea-Dwellers to Rolex with a concise but encouraging report on their performance.

"I am, this date [21 April 1969] sending you the two experimental 'Sea-Dweller' chronometers with helium-valves that you loaned the sealab project. As you are no doubt aware, the at-sea phase was terminated due to both a personnel casualty and material problems.

"Watch #1 was taken to a depth of 602 feet in the open sea and was then decompressed in a helium-oxygen environment. It kept correct time during the eight-day dive period. Watch #2 was also worn during the same dive and it too kept good time.

"The only unfavourable comments from the aquanauts are: (1) the snap on the watch band is rather bulky and springs open from the rear of the clip; (2) the watch crystals appear to be subject to scratching and scar very easily; and (3) the bezel has a tendency to slip. Trusting you will find this information of value."[6]

Carpenter retired from the navy the same year. He founded Sea Sciences, a subaquatic consultancy company, worked with Jacques-Yves Cousteau, and remained a lifelong friend of Rolex. Interviewed in 2008, five years before his death, he said, "the Rolex Submariner and Rolex Sea-Dweller have a cachet that is unmatched by any other watch. The Rolex brand is the most highly respected in my book. The Rolex Submariner was the preferred diving watch for U.S. Navy Sealab."[7]

Almost a decade after its triumphant descent into the Challenger Deep, the U.S. had been forced to retreat from the underwater frontier, but the Sea-Dweller was by now established as a deep-sea essential.

ASA
CONTROL CENTER

SEALAB

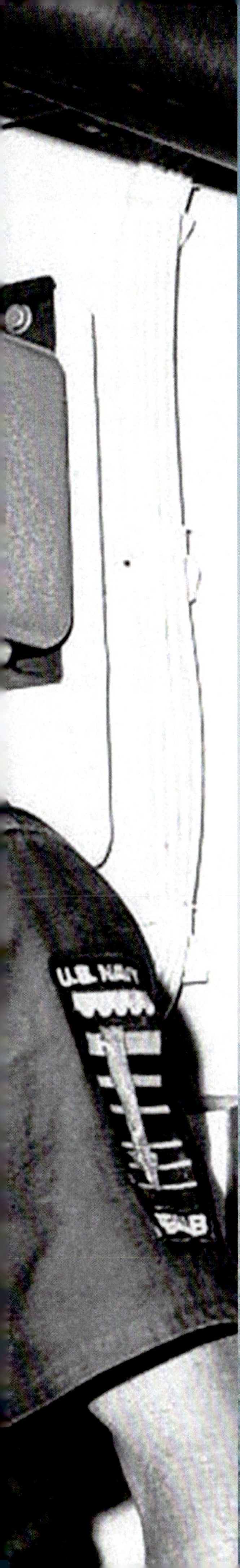

U. S. Navy Sealab III being lowered into the Pacific Ocean off the coast of California, in February 1969.

Opposite: Scott Carpenter inside Sealab II.

This special version of the Sea-Dweller, Ref. 1665, 40 mm, worn by Joe MacInnis, bears the words "SUBMARINER 500 M – 1650 FT" on its dial – an inscription that made the watch sought after by collectors.

Opposite, from top: A letter from Scott Carpenter to René-Paul Jeanneret dated 29 November 1967 and signed by Jeanneret on 7 December 1967; Carpenter wearing a Rolex watch, circa late 1960s.

"AN IMPORTANT NATIONAL FRONTIER LIES JUST UNDER THE SURFACE OF THE SEA, AND WE MUST GET DOWN TO IT AND STAY THERE."

SCOTT CARPENTER

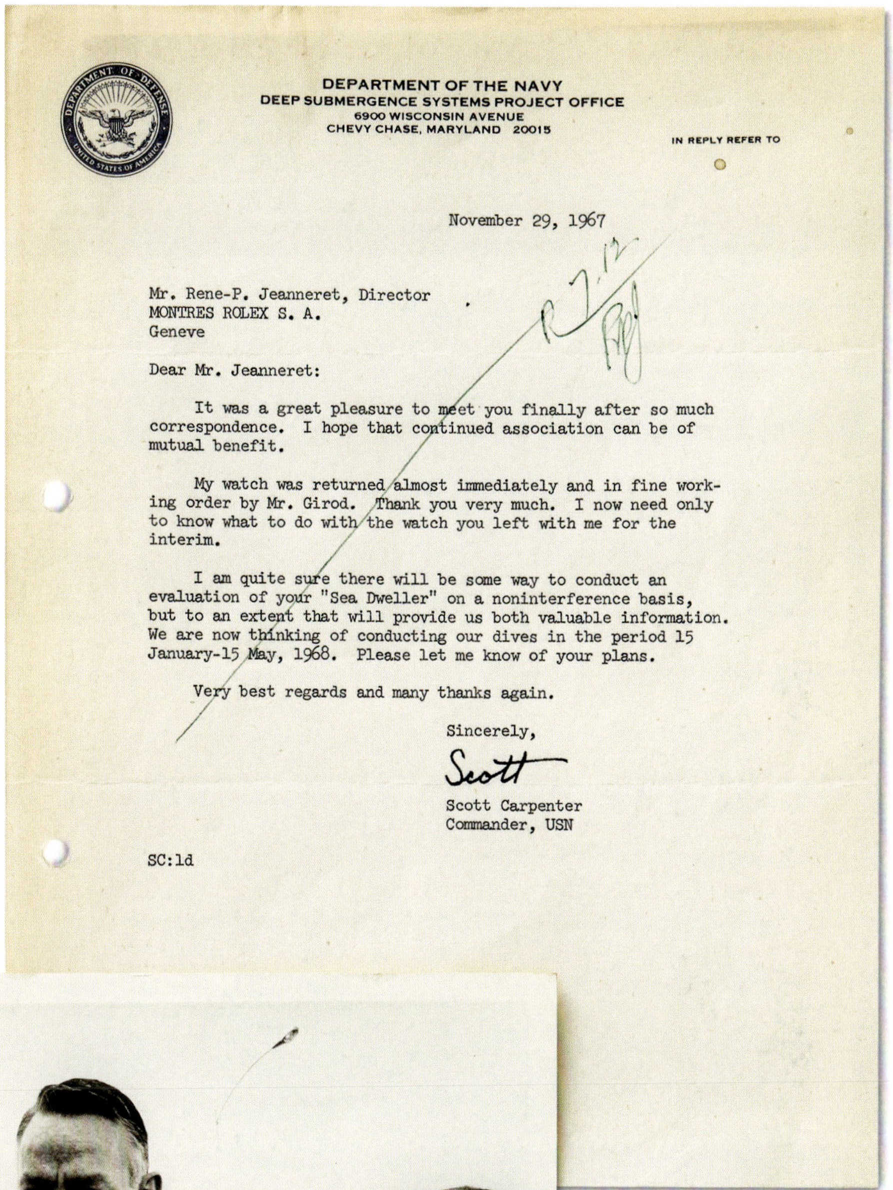

DEPARTMENT OF THE NAVY
DEEP SUBMERGENCE SYSTEMS PROJECT OFFICE
6900 WISCONSIN AVENUE
CHEVY CHASE, MARYLAND 20015

IN REPLY REFER TO

November 29, 1967

Mr. Rene-P. Jeanneret, Director
MONTRES ROLEX S. A.
Geneve

Dear Mr. Jeanneret:

It was a great pleasure to meet you finally after so much correspondence. I hope that continued association can be of mutual benefit.

My watch was returned almost immediately and in fine working order by Mr. Girod. Thank you very much. I now need only to know what to do with the watch you left with me for the interim.

I am quite sure there will be some way to conduct an evaluation of your "Sea Dweller" on a noninterference basis, but to an extent that will provide us both valuable information. We are now thinking of conducting our dives in the period 15 January-15 May, 1968. Please let me know of your plans.

Very best regards and many thanks again.

Sincerely,

Scott

Scott Carpenter
Commander, USN

SC:ld

Joe MacInnis, 1978.

JOE MACINNIS

Rolex Testimonee

Even by the exalted standards of the Testimonees who have appeared in Rolex advertising, Joe MacInnis is a remarkable man. A medical doctor, pioneer diver, explorer, journalist, underwater habitat designer, broadcaster, motivational speaker, author, government advisor, campaigner and poet, this extraordinary individual first came to the attention of Rolex on 21 July 1967 when Victor R. Royce, an employee of Rolex Canada, read an article about MacInnis and wrote to Rolex Geneva.

"You will notice that the divers do reach a depth up to 1,250 feet and we were wondering how this would effect [sic] the performance of a Rolex Submariner, for example, which is only guaranteed waterproof to 660 feet. We would obviously not wish to involve Rolex in something beyond which it was considered capable. We should like to have the answer to this question before we contact Joe MacInnis."[1]

As soon as the letter crossed his desk, René-Paul Jeanneret spotted the potential for the sort of combined research, development and publicity activity at which he excelled. Dive depths of 1,250 feet might exceed the capabilities of the regular Submariner, but would be perfect for testing the incipient Sea-Dweller, as he wrote to Royce: "We shall see what can be done about presenting Joe MacInnis with one of our new Submariner Sea-Dweller models."[2] Jeanneret also asked Rolex New York to find out a little more about the intriguing Joe MacInnis.

Born in 1937, MacInnis took his first dive at age 18. He was at a swimming competition in Fort Lauderdale, Florida, "when someone suggested we explore the near shore reef. I got checked out in the pool in the morning and made a 60-foot scuba dive in the afternoon."[3] By his own admission an indifferent medical student – "I was interested in other things"[4] – he nevertheless impressed Edwin Link, the flight simulation pioneer, entrepreneur and inventor who, by the early 1960s, was becoming interested in deep-sea diving and underwater habitation. Link saw something in the young medical graduate, and in 1963 elected to give him a scholarship to study diving medicine and work with U.S. Navy divers preparing deep-dive oxygen and carbon dioxide monitors.

"WE CAN NOW LIVE BENEATH THE SEA AND IN THE VACUUM OF SPACE – AND ARE THUS IN A POSITION TO BEGIN AN EXTENSIVE EXPLORATION OF BOTH."

JOE MACINNIS

Link's hunch had been correct. Within four years, MacInnis had become a highly sought-after medical consultant to the diving world. In 1966, he was part of a team that retrieved an H-bomb that was missing off the coast of Spain following a B-52 crash. He was also working with Ocean Systems, involved in the development of equipment enabling industrial divers to work at remarkable depths. In summer 1967, Ocean Systems divers had descended to 636 feet, a feat that encouraged the industrial diving business to believe that divers could "work freely at least to 1,000 feet." The ramifications were huge: "All the world's offshore continental shelves, with a total area as big as Africa and a 1,000-foot maximum depth, were suddenly open to exploration and commercial exploitation."[5] MacInnis was also involved in the much anticipated Sealab III.

It was inevitable that news of his subaqueous exploits would reach Rolex sooner or later.

On 13 May 1968, MacInnis wrote to Rolex:

"This letter is to thank you sincerely for the Rolex Sea-Dweller. It is a superb timepiece and will be most valuable in our work. I will endeavour to obtain photographs of the watch 'in action' during our upcoming laboratory and open-sea dives.

"Again, my sincere thanks, I am most honoured by the present."[6]

This otherwise formal typed letter comes alive with a hasty, enthusiastic postscript scribbled in pencil in untidy capital letters:

"THE WATCH HAD ITS FIRST DIVE THIS WEEKEND.
SOON 1,200 FEET. AGAIN MY THANKS, JOE"[7]

Forever thereafter, the fates of Joe MacInnis and Rolex would be intertwined.

Within a year, MacInnis had worn his Sea-Dweller during Sealab III training in San Francisco; in Jacques-Yves Cousteau's laboratory, where he undertook recondite sounding experiments;[8] in waters off Caracas, where he was involved in the search and recovery of a Pan Am Boeing 707 that had crashed in 1968, making "twenty-one dives to depths of 370 feet"[9] described as "the deepest salvage task that has ever been undertaken";[10] and in British Honduras (now Belize) for a Smithsonian shark study programme.

He had also been issued another Rolex.

In March 1970, MacInnis provided another update on the active life of his Sea-Dweller. It had been worn while "conducting under-the-ice studies in the Great Lakes,

An advertisement announcing Joe MacInnis' dive, 1974.

Rolex went under at the North Pole.
In April of 1974, a team of professional underwater experts, directed by Dr. Joe MacInnis, dove beneath the Polar Ice Cap at the geographic North Pole. Expertise, confidence and precision were the orders of the mission.
Both the divers and their Rolex watches performed flawlessly. But you'd expect nothing less from a Rolex.
In an age of obsolescence and gimmickry, this simple classic virtue of a Rolex is indeed a rarity.
The Rolex Submariner. Oyster case tested pressure-proof down to 660 feet. Officially certified, self-winding superlative chronometer. In stainless steel or 18 kt. gold with matching bracelets.
Write for brochure: Rolex Watch Company of Canada Ltd., 80 Richmond Street West, Toronto, Ontario M5H 2B8.
ROLEX

Joe MacInnis was the first person to explore the ocean beneath the North Pole.

where we have been diving to the SUBLIMNOS habitat"[11] – an underwater module, designed, funded and built by MacInnis for extended underwater habitation and the first to remain functional for a year. "This means that the watch has been exposed to the extreme heat of the warm-up cabin and the extremely cold water found beneath the ice. The watch performed flawlessly ... and this is especially important since much of our scientific diving involves much needed time accuracy."[12]

In addition to the Pan Am salvage in Venezuelan waters, it had been "worn beneath the cool green waters of the Gulf of Mexico, where another team of Ocean Systems, Inc., divers conducted the deepest salvage yet carried out by man ... at 490 feet. It was work carried out for the U.S. Air Force and U.S. Navy ... and the Rolex was critical in the timing of the bottom phases and decompression aspects of these very deep exposures.[13]

"In July 1969, I was a member of a team that conducted a 1,000-foot human experiment ... Our objective was to evaluate how rapidly humans could accommodate to these tremendous pressures. The Rolex was worn by one of the divers and functioned perfectly during the rapid pressure build up. The breathing mixture was made up of oxygen and helium, and at no time did the helium, a light and highly diffusable gas, affect the Rolex's function. This is probably due to the unique helium vent valve ...

"My 'Sea-Dweller' has been beneath the ice ... in small submersibles and submarines ... and at all times has worked perfectly. It has been a good friend ... and I wanted you to know about it."[14]

When a copy of this letter arrived in Geneva, Jeanneret was ecstatic, scribbling on the cover letter in green ink "this is by far the best testimonial on the Sea-Dweller so far." For good measure, block capitals, circled and written in red ink, read "MOST IMPORTANT TESTIMONIAL!"[15]

It is interesting to note that MacInnis' letter was written on the headed paper of Undersea Research, which was also printed with a few lines by Icelandic poet Davíð Stefánsson:

"THE EXPLORER
IS THE POET OF ACTION
AND EXPLORATION
IS THE POETRY OF DEEDS"[16]

This quotation shows how MacInnis saw himself as more than just a practised diver and medical professional.

Increasingly, his activities were wrapped in a cloak of hierophanic mysticism: a sense of wonder, poetry and enlightenment; an awareness of the fragility of environments he was experiencing; and a highly developed awareness of man's place in the cosmos overlaid his scientific work. He combined an almost Wordsworthian sensitivity to the natural world with the curiosity of a pioneer explorer, the tireless zeal of a committed environmental campaigner and the rigour of a late-20th-century man of science.

Like others at the time, including Scott Carpenter and Jacques-Yves Cousteau, MacInnis was one of the charismatic, messianic visionaries of the 1960s and 1970s who saw a future for humankind under the waves. His ideas conflated the excitement of the Space Race with an environmental concern decades ahead of its time and a spiritual awakening, as the foreword to *Underwater Images*, a book of his poetry coupled with his underwater photographs, demonstrates:

"The age of Aquarius. A time of splendid adventure. A time when man stepped off the earth – into space and into the sea. An incredible decade, when we placed the first footprints on the moon and began the elegant return journey to the oceans where life began.

"The wings of Apollo have given us a new view of our old planet. We are now able to fully realize the fragile roundness of our space-ship earth, and within this roundness visualize the past and present of man. Until now, all that man was and is took place inside the narrow envelope of air which surrounds us, but today's authoritative access to space and to the sea has dramatically altered this situation. We can now live beneath the sea and in the vacuum of space – and are thus in a position to begin an extensive exploration of both. The wings of Apollo allow us to see a brightly lit sphere, orbiting a delicate ellipse against the black womb of the universe. We have at last discovered the magnificence of the journey – a horizonless orbit – hinged with potential."[17]

In 1970, he established the MacInnis Foundation, which set out a number of ambitious goals that were, in some cases, idealistic to the point of being utopian:

"To identify and solve problems associated with improving man/environment harmony.

"To enhance manned underwater exploration by improving and utilizing underwater submersibles, habitats and other life-support equipment.

"To encourage improved understanding of the physical and chemical, biological, and geological aspects of the underwater environment.

"To use the new-found access and understanding of the underwater environment as an educational vehicle to enhance creative and learning processes.

"To encourage high school and university students to use the underwater world and underwater habitats as 'classrooms' of the new frontier.

"To disseminate new information as rapidly as possible into the mainstreams of Canadian and world cultures, including public, private and government agencies.

"To conduct underwater research which bears direction on world-man's current problems including: population and pollution excess and protein deficiency."[18]

It is striking to note how much of this 1970 mission statement finds an echo in Rolex's Perpetual Planet Initiative over half a century later.

As well as supplying him with watches, Rolex soon became a financial supporter of MacInnis' work. The archives are full of correspondence relating to his activities.

MacInnis was still near the beginning of a long and productive work life when, in 1969, he met Pierre Trudeau, the charismatic Canadian Prime Minister who dominated the country's political life from the late 1960s until the mid-1980s. The two men would dive together many times, and MacInnis would go on to become a government advisor, helping to draft the country's first ocean policy. He would also lead 10 government-supported expeditions under the Arctic Ocean. He was the first person to dive in the icy waters beneath the North Pole. His adventures, whether hunting shipwrecks under Arctic ice or navigating the Northwest Passage, captured the public imagination so vividly that they became part of Rolex's advertising. He was among the first to dive to the wreck of the Titanic, and when James Cameron and Rolex dived to the bottom of the Mariana Trench, he served as both medical advisor and chronicler. Five-and-a-half decades after he put that early Sea-Dweller on his wrist, Rolex continues to be part of MacInnis' story, and the Rolex logo is prominently displayed on MacInnis' website.

The story of its lifelong relationship with MacInnis is a history in miniature of Rolex's growing awareness of the precious fragility of the environment that the Submariner has done so much to open up to humankind.

Alain Jourde and Claude Bourdier set a world record at a simulated depth of 610 metres (2,000 feet) at the Comex facility in Marseille.

UNDER PRESSURE

In June 1969, as Sealab was being wound up, a suave, tanned Henri Delauze invited television cameras into his new hyperbaric test centre in the south of France. The 1960s had been boom times for Delauze. By the decade's end, Comex was particularly active in the North Sea, and in 1975, Comex's UK subsidiary was "contracted to 18 exploration rigs and employed on five major construction projects, supported by 200 deep divers at any one time and 50 shore-based personnel."[1] Dives of up to 245 metres (804 feet) were routinely undertaken by the company's most experienced divers and its specialist equipment was in high demand. Comex's life-support complexes – deck decompression chambers, transfer-under-pressure capsules and bells – had already found a ready market with other companies in the North Sea and across the world. Owners and operators of three saturation complexes themselves in 1973, Comex UK purchased a further six in 1974.[2]

Wherever offshore oil exploration was big, so was Comex and the nearly homonymous Rolex. When Offshore Europe '77, "Europe's largest exhibition of offshore oil and gas recovery equipment which attracted 500 exhibitors"[3] took place in Aberdeen, Rolex staged a special exhibition of its watches at celebrated local jeweller Jamieson & Carry, promoted with evening receptions and advertising in *Roustabout*, a local oil trade publication, Aberdeen's two newspapers, and even on television.

For those who got it right, the oil rush of the 1970s was a bonanza. Having brought Comex to a leading position in offshore oil exploration and drilling, Delauze was determined to maintain domination of the field through advanced research into all aspects of diving, with the aim of bringing men to ever-greater depths. He was particularly proud of a hangar-like space at his headquarters that housed a network of seven interconnected hyperbaric chambers. One chamber, he claimed, was the biggest in the world. He talked of eventually bringing divers to incredible depths of 600 to 700 metres (1,969 to 2,297 feet) and developing breathing mixtures that relied not on helium, but on highly explosive hydrogen, which was half its weight.

Spring 1972 witnessed a landmark event in both the history of saturation diving and high-pressure horology when two Comex divers, Patrice Chemin and Robert Gauret, set a world record for a "dive" at a simulated depth of 610 metres (2,000 feet) at the Comex facility in Marseille, sealing one of the most celebrated professional partnerships in the history of Rolex. "For 18 days, they worked in the extreme conditions experienced at this enormous depth, and for 18 days their Rolex Sea-Dweller watches worked perfectly," proclaimed the advertisement placed by Rolex to celebrate this achievement. "Comex were so pleased with the Sea-Dwellers' performance that they at once ruled that only Rolex diving watches and chronometers are to be used by all their divers in the future."[4] Just to reinforce the message of the advertisement, Chemin was quoted as saying "The Rolex Sea-Dweller alone is able to give the exact time at this depth."[5]

For the next 25 years, Rolex advertising would periodically highlight its contribution to Comex's staggering achievements. During the 1970s, the Submariner (Ref. 5514) was a Comex-specific watch with a helium escape valve.

Rolex advertisement, detailing the Comex Physalie experiment, 1970s.

Opposite: Submariner, 1974, Ref. 5514, 40 mm, stainless steel, waterproof to 200 metres (660 feet). This watch was used by Comex, whose name is inscribed on the dial.

Rolex Sea-Dwellers were worn again on another 600-metre (1,967 feet) simulated dive in 1974. The brand was also an integral component of Comex Special Project Janus IV, when a team of Comex divers was sent to the bottom of the Mediterranean off the coast of Cavalaire-sur-Mer, where they lived and worked for several days at depths of up to 501 metres (1,644 feet), undertaking pipeline connection duties. The operation was minutely calculated using new compression tables, while the controlled administration of nitrogen to reduce high-pressure nervous syndrome (involuntary movement experienced by deep divers breathing heliox) doubled the accepted depth at which diving teams could work. But for Comex and Rolex, this remarkable depth was a way station rather than a destination.

"Now Comex are involved in solving the difficult problems of even deeper descents by their divers, but they remain confident of overcoming them.

"One thing is certain.

"Wherever a Comex diver goes, his Rolex goes with him."[6]

Delauze was confident that by the 1980s, humans could be sent to depths of 1,000 metres (3,281 feet); the chief difficulty was that heliox could only be used to 700 metres (2,297 feet). On 1 July 1983, Delauze himself took a five-minute dive of "only" 91 metres (299 feet). Its brevity belied its significance: Delauze was breathing hydrox, a potentially explosive hydrogen/oxygen mixture. The way to one kilometre underwater was theoretically clear, an achievement Rolex welcomed with the ironic strapline "One day soon, the Comex divers will be able to go as deep as Rolex."[7]

In 1978, the Sea-Dweller Ref. 1665 had been replaced by the first sapphire-crystal-equipped Sea-Dweller Ref. 16660; the new watch doubled the effective operating depth of the Sea-Dweller to 1,220 metres (4,000 feet). This exceptional water resistance was a source of much pride and some slight gloating: "For Comex divers, 1,220 metres below the sea is yet another frontier," explained the advertising of the 1980s, adding smugly, "for Rolex it has long been possible."[8] As it happened, the deepest a Comex diver would ever go was 701 metres (2,300 feet), a record established by Theo Mavrostomos in a hyperbaric chamber in 1992 breathing a hydro-heliox mixture and requiring 24 days of decompression. Although no such depth was ever achieved by a Comex diver in the ocean, the Hydra VIII project of 1988 saw divers working at a depth of 534 metres (1,752 feet) off the French coast near Marseille.

ROLEX
comex
SUBMARINER

FOR COMEX DIVERS, 1,220 METRES BELOW THE SEA IS YET ANOTHER FRONTIER, FOR ROLEX IT HAS LONG BEEN POSSIBLE.

A selection of Rolex watches worn by Comex divers *(from top)*:
Submariner Date, 1979, Ref. 1680, waterproof to 200 metres (660 feet);
Submariner Date, 1992, Ref. 16610, waterproof to 300 metres (1,000 feet);
Submariner Date, 1982, Ref. 16800, waterproof to 300 metres (1,000 feet);
Sea-Dweller, 1982, Ref. 16660, waterproof to 1,220 metres (4,000 feet).

ROLEX
comex
1000ft=300m
SUBMARINER

In the 1960s, Canadian company HYCO became renowned for its development of the *Pisces* submersibles, which were designed for use in ocean research and exploration.

FROM HYCO TO THE HYDROGRAPHIC DEPARTMENT

Comex remains by far the best-known of Rolex's subaquatic associations, its fame reflected in the desirability of Comex-signed Rolex watches amongst 21st-century collectors, but it was not the only such company to provide pioneering input into the development of the Sea-Dweller.

"Organised in 1964. Specialised in the design and construction of systems required to accomplish work tasks in the sea – effective work in the water – ocean engineering or marine construction – designed and manufactured more deep diving manned submersibles in active use today than any other company."[1] Preserved in notes dating from the early 1970s held in the Rolex Archives, this is the admirably concise, almost telegraphic description of the Vancouver-headquartered International Hydrodynamics Co. Ltd.

Mercifully known by the less cumbersome acronym HYCO, the Canadian company was famous for its *Pisces* submersibles. *Pisces I* was built in 1965 for a crew of two operated at a maximum depth of 366 metres (1,200 feet); three years later, *Pisces II* was launched, a vessel capable of taking its crew of three to 793 metres (2,600 feet); the following year saw *Pisces III* add another 305 metres (1,000 feet) to the maximum operating depth; and in 1971, *Pisces IV* was launched with a remarkable 2,000 metres (6,500 feet) of depth capability.

Of course, the *Trieste* had reached far greater dive depths, but that had been a piece of state-funded, military-scientific Cold War one-upmanship; by contrast, HYCO was a business offering a commercial product created to meet a specific real-world need. The Alan Trice-designed *Pisces* manned submersibles were made for use in submarine research and exploration.

Although the firm went out of business in the late 1970s, for a period of about a decade, HYCO represented the sharpest point on the cutting edge of manned submersible technology. As *Maclean's* magazine put it in 1969: "*Pisces* subs are second to none in their reputation for outstanding work in oceanographical research, fisheries studies, undersea oil exploration and geographical surveys."[2]

Specializing in extreme-depth diving and working under polar ice, as well as manufacturing, selling and leasing submersibles and scientific vessels, HYCO seemed to be a one-stop shop for deepwater diving needs. The company also offered design services and expertise in pollution control, "cold-water diving, pipeline installation, explosives welding and burning, [and] cable and pipeline inspection."[3]

In November 1967, knowing of his interest in diving and underwater research, Roland Puton of Rolex Canada wrote to Jeanneret about HYCO's *Pisces* programme. Jeanneret sensed an opportunity and responded immediately.

"A 'DEEP-SEA' MODEL WILL SURELY FIND A BIG MARKET EVERYWHERE."

HANS WILSDORF, *Founder of Rolex*

"I wonder if you have heard about a model we are developing at this time and which will be called the Sea Dweller. This is a modified version of the Submariner. It will have a date (without a cyclope lens) and will be waterproof and pressure-proof down to 610 metres (2,000 feet). We are working on the first series of prototypes and some time early next year, you will hear more about this.

"Until this model is officially launched, we are interested in having a series of practical tests carried out as soon as further, tested prototypes become available, and it would be a great advantage to be in touch with as many divers as possible who go down very deep – aided by the latest mixtures of gas – and who stay under water for a considerable length of time.

"You may want to get in touch with Messrs. Thomson, Sorte and Trice and discuss our project with them.* If they are willing to carry out a similar kind of test, you might work out some scheme to get maximum publicity out of it on a Canadian basis. No doubt an experiment carried out in the waters near Vancouver will have greater impact locally than a similar experiment carried out elsewhere, even if the latter is a more spectacular one."[4]

Thomson, Sorte and Trice were only too happy to pursue the association, which came at a crucial time for Rolex's continuous efforts to take precision timekeeping ever deeper beneath the waves. By February 1968, Jeanneret was in touch with Donald Sorte directly, informing him that Rolex was "bringing out a new type of divers' watch destined for very great depths" and that two prototypes would be flown out to Rolex Toronto. Describing these as "Mark II" Sea-Dwellers, he explained that Rolex had "improved the model until we could guarantee it to 2,000 feet (actually, it has been tested as far as 3,000 feet) and kept it under test at those depths for over one month." His letter offers further intriguing insights into the pace of development when he writes, "although we do not yet have a dial with marking for the final model and the inscription on the watches says 500 metres – about 1,650 feet – these watches are guaranteed to be capable of withstanding the pressure at 2,000 feet plus a large safety margin, for long periods of time."[5]

HYCO was invited to subject the new Mark II Sea-Dwellers "to any kind of test you deem necessary or interesting, regardless of the fact that the Sea-Dwellers may be damaged in the process. All we are interested in is to find out what this watch will or will not do in practical applications: we should be most grateful if you would keep us posted as to its performance, voicing any comment or criticism which might help us improve the model and bring out a final version which would be absolutely perfect for the present and future generations of deep-sea dwellers."[6]

Following an inspection dive in March 1968 at the Nanoose Bay torpedo range in the Georgia Strait, initial comments were favourable. Photographs of the Sea-Dweller, attached to the outside of the *Pisces I*, were sent to Rolex: "The third picture shows the watch at a submerged depth of 1,350 feet. Note the depth gauge on the left and at right, the outside pressure gauge which registers approximately 600 psi. We were at this depth for four hours. During all phases of testing, the watch performed beautifully."[7]

Ties between HYCO and Rolex were strengthened that summer with a visit to Geneva during which Jeanneret mentioned the patent application from November 1967 and the development of a new series of Sea-Dweller called the "Mark IV", equipped with the helium escape valve that would come to define the watch. In August 1968, HYCO put a Sea-Dweller to the test 610 metres (2,000 feet) beneath the ice in waters off Greenland on a submersible that they had painted with the Rolex crown.

At the beginning of the 1970s, the Sea-Dwellers were still in service on the wrists of Trice and Sorte, and in 1971, Jeanneret sent some more up-to-date watches. "These Sea-Dwellers will be equipped not only with a helium valve, but also with our latest model winding crown, the experimental Triplock, which is waterproof even when unscrewed down to at least 80 meters."[8]

In March 1971, news broke that HYCO had just signed a historic deal to sell a *Pisces* submersible with a depth capability of 1,829 metres (6,000 feet) to the USSR. This sale generated an intriguing story linking Rolex with Soviet scientists, related at length in correspondence from Rolex Canada to Geneva.

* Mack Thomson, Donald Sorte and Alan Trice, founders of HYCO.

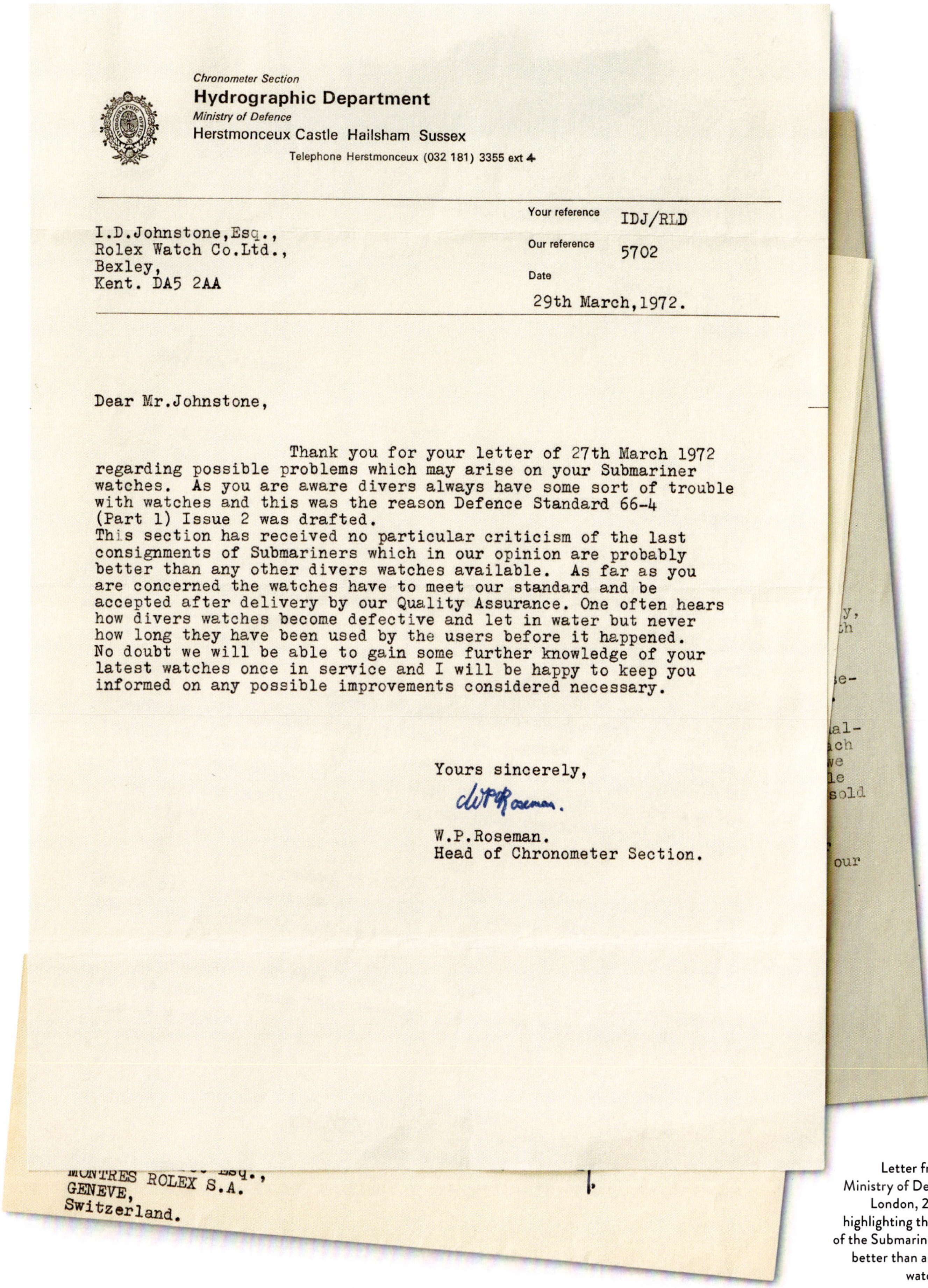

Chronometer Section

Hydrographic Department

Ministry of Defence

Herstmonceux Castle Hailsham Sussex

Telephone Herstmonceux (032 181) 3355 ext 4

Your reference IDJ/RLD

Our reference 5702

Date 29th March, 1972.

I.D.Johnstone,Esq.,
Rolex Watch Co.Ltd.,
Bexley,
Kent. DA5 2AA

Dear Mr.Johnstone,

Thank you for your letter of 27th March 1972 regarding possible problems which may arise on your Submariner watches. As you are aware divers always have some sort of trouble with watches and this was the reason Defence Standard 66-4 (Part 1) Issue 2 was drafted.
This section has received no particular criticism of the last consignments of Submariners which in our opinion are probably better than any other divers watches available. As far as you are concerned the watches have to meet our standard and be accepted after delivery by our Quality Assurance. One often hears how divers watches become defective and let in water but never how long they have been used by the users before it happened. No doubt we will be able to gain some further knowledge of your latest watches once in service and I will be happy to keep you informed on any possible improvements considered necessary.

Yours sincerely,

W.P.Roseman.

W.P.Roseman.
Head of Chronometer Section.

Letter from the British Ministry of Defence to Rolex London, 29 March 1972, highlighting the performance of the Submariner as "probably better than any other divers watches available".

Pisces

Sea-Dweller 2000;
this watch was tested by the
Admiralty Experimental
Diving Unit in the 1970s.
Ref. 1665, 40 mm, stainless steel,
waterproof to 610 metres
(2,000 feet).

Opposite: Pisces I enters the water,
circa late 1960s. At the time,
Pisces crafts represented
the cutting edge of
manned submersible
technology.

Pisces I, displaying the Rolex crown on the side of the hull, is loaded onto a Canadian Air Force plane, circa late 1960s.

"Dr Igor E. Mikhaltzev, Chief Oceanographer, Institute of Oceanology, Soviet Academy of Sciences, is the scientist with whom this contract was negotiated. According to Mr Sorte, Dr Mikhaltzev is the foremost oceanographer in the Soviet Union and one of the top three in the world."[9] As a mark of his esteem, Sorte had decided to present the Soviet scientist with an 18 ct gold Rolex Thunderbird with a matching Jubilee bracelet that he had bought a few years earlier. "The Thunderbird is now at our office for a complete service and refinishing job, so that it looks as new. Since the watch presumably belongs to Mr Sorte, we see no reason why this presentation could not take place.

"In addition, Mr Sorte would like to purchase from Rolex four additional Sea-Dwellers with helium valve. These watches would be intended for the four Soviet scientists who will spend about three months in the Vancouver area with International Hydrodynamics on a special training program. These four scientists are scheduled to arrive in Vancouver imminently."[10]

This act of underwater détente was potentially a big news story.

"The first *Pisces* had a large Rolex logo and crown painted on the exterior. Mr Sorte would like to paint a similar logo on the *Pisces* ordered by Dr Mikhaltzev. The interest here is that the loading of the *Pisces* onto the gigantic Russian vessel that is expected to take it to the Soviet Union from Vancouver should give Canadian news media a field day, in addition to which the use of the *Pisces* will not be restricted to Russia, but also may well be used by other Iron Curtain countries at appropriate times."[11]

Rolex Geneva agreed that "it would be nice if the *Pisces* could have the Rolex logo and crown painted on its exterior on its trip to the USSR."[12] But, as it turned out, with the Cold War at its height, the sale of the *Pisces IV* to the Soviet Union was cancelled and the vessel was purchased instead by the Canadian government.

Meanwhile, the new Sea-Dwellers Jeanneret had sent were reported to be "performing very well and keeping accurate time" – so accurate, in fact, that the "SEA-DWELLER, worn by Mr Alan R. Trice, Director of Field Operations, was used as a back-up check on our Satellite Navigation System, which periodically was inoperative or gave the incorrect time. The accuracy of the SEA-DWELLER assisted in determining when the navigation system was faulty. Mr Trice has expressed to us his complete satisfaction with the performance and reliability of his Rolex."[13]

As well as complete satisfaction, it is also possible to detect slight envy of the symbiotic relationship between Comex and Rolex. HYCO was keen to benefit from being part of Rolex's advertising as shown in a request to supply watches for use during fulfilment of a contract to lay a transatlantic cable. "It would please them very much to place a Sea-Dweller on the 'wrist' of their submarine*, and to take pictures which they would let us have for advertising purposes. Apparently, the only thing they want to achieve is that we use their pictures or story or both, mentioning HYCO, for our advertising. If we do not want to use any material they submit – at no charge – there will be no hard feelings. All they expect us to do is to supply two Sea-Dwellers and to be responsible for any damage to these."[14]

Once again, Rolex authorized the testing of the watches until destruction, but as it happened, it was the mission rather than the watches that came to grief. The cable lifting *Pisces III* to the surface snapped, letting the sub and its two occupants fall to the seabed. The dramatic search and eventual rescue, effected after 72 hours of mishaps, bad weather and worse luck that had almost ended the lives of the two submariners, made headlines around the world; even amidst the tension during those action-packed three days, HYCO made it clear to Rolex that the *Pisces V*, which had been involved in the rescue, had worn a Sea-Dweller on its "wrist".

HYCO was one of the most technically innovative of Rolex's partners, and the beginning of the association coincided almost exactly with the launch of the first Sea-Dweller 2000. However, as was customary at Rolex, HYCO was far from the only organization to be testing this new watch in the field. In addition to HYCO and Comex, Rolex turned to an organization whose association with the subaqueous capabilities of Rolex watches predated the launch of the Submariner, let alone the Sea-Dweller.

In the spring of 1972, Rolex made contact with W. P. Roseman, Head of the Royal Navy's Chronometer Section. It had been 20 years since the Royal Navy divers had taken their first dives with Rolex watches in the chilly waters of Loch Fyne, and Rolex had been supplying the Royal Navy since 1955. In the late 1950s and early 1960s, Rolex advertising in British newspapers used the strapline "The Royal Navy dives with Rolex" above a photograph of a Royal Navy diver wearing a Rolex.[15]

"There has been considerable cooperation between Rolex and the Admiralty for the past 15/20 years in connection with deep sea diving watches," wrote Rolex's London Sales Director in 1972. "Since then, we have supplied Rolex Oyster Perpetual Submariners in quantity to the Admiralty as well as, of course, to many other navies throughout the world."[16]

Although the Royal Navy no longer appeared in its advertising, Rolex remained in regular contact regarding supply and maintenance of watches. Indeed, it was in response to a routine letter from Rolex's UK office "regarding possible problems" that the Submariner might experience that Roseman had written back. "This section has received no particular criticism of the last consignments of Submariners which, in our opinion, are probably better than any other divers' watches available."[17]

Anyone who knew him, knew that this was Salzburg-Festival-level orchestral music to Jeanneret's ears, and even though he was shortly to retire from the company to which he had dedicated over quarter of a century of his working life, he could smell a promotional coup all the way from Geneva. Describing Roseman's observation as "highly gratifying",[18] he wrote excitedly to Rolex in Britain.

More than half a century later, his enthusiasm bounds off the flimsy carbon copy on which his words are preserved in the Rolex Archives.

"That an official representative of the Admiralty should call our Submariners 'in our opinion, probably better than any other divers watches available' is highly gratifying; do you think we have any chance whatsoever of obtaining a release?

"Please try to phone Mr Roseman and put out a very tentative feeler, withdrawing it immediately if you meet the slightest resistance. If you could manage to get his permission – or that of his responsible superiors – for us to use the above words in an ad, saying that they came from the Admiralty's Chronometer Section or whatever the correct title would be, that would indeed be success.

"I greatly look forward to your news and would like to congratulate you and to thank you for taking intelligent action (as usual)."[19]

After further exchanges of letters, I. D. Johnstone of Rolex UK "thought it was about time that personal contact was once again made with Mr Roseman, and I,

* An external articulated manipulator arm.

INTERNATIONAL HYDRODYNAMICS
HYCO

therefore, took the opportunity of taking the recent order of 57 Submariners down to him personally on Tuesday, 2nd May."[20]

It was, as he put it, "a most interesting and enlightening visit".[21] Johnstone was particularly impressed by the lengths to which Roseman's department went to ensure the water resistance of watches.

"To start with, they regard the Etancheiscope principle as not being a satisfactory test as to whether or not a watch is waterproof.

"First of all, the watches are put into a sealed cabinet from which the air is completely removed. The case is then filled with dry nitrogen. The watches are left in this condition with the backs off for 24 hours in order to become completely saturated in the nitrogen to ensure a humidity factor of less than 20 per cent. The watches are then sealed up, still inside the cabinet, so that no air can possibly get inside the watch. They then go into an oven, which is heated up to 35 degrees Celsius. Having been removed from the oven, a piece of ice is placed on top of the glass and if no condensation appears they are then put into a fathometer type instrument for at least 6 hours at a depth of 120 meters. The oven and ice treatment is then repeated. If no condensation appears, then they are satisfied that the watch is in fact waterproof.

"It was very interesting to be told that many of the watches supplied to the Admiralty many years ago are still surviving this test without problem. They are obviously well aware that the performance of a watch such as the Submariner is absolutely vital to a diver whose life may depend upon it."

Johnstone clearly warranted Jeanneret's little encomium as, in addition to personally delivering the order of 57 Submariners and following up the advertising enquiry, he had brought some new products with him that interested Mr Roseman.

"I also took the opportunity of showing him a Sea-Dweller, in which he was very interested. As you are aware, we have eight Sea-Dwellers with Mr Common at HMS *Vernon* at Portsmouth, and Mr Roseman will be contacting him with a view to getting hold of one or two of them which he can test in his own way. I also have a feeling that the next enquiry from the Admiralty will be for the 5513 Submariner, but with the helium valve in it."[22]

A technician works on the manipulating arm of a HYCO Submarine. Both wear matching Rolex watches.

"THE PERFORMANCE OF A WATCH SUCH AS THE SUBMARINER IS ABSOLUTELY VITAL TO A DIVER WHOSE LIFE MAY DEPEND UPON IT."

I.D. JOHNSTONE

Sea-Dweller 4000, 1978, Ref. 16660, 40 mm, stainless steel, waterproof to 1,220 metres (4,000 feet).

Opposite: An illustration from a HYCO Enterprise brochure depicting its divers at work, circa 1967.

"Bond. James Bond." Scottish actor Sean Connery portrays Ian Fleming's famous superspy in a publicity shot for 007's screen debut in *Dr. No* (1962).

SECRET AGENT, FAMOUS WATCH

It is nighttime in a small harbour. A diver emerges from the water. Checking quickly to ensure that he is not observed, he rips off his breathing gear.

In an instant, he is out of the inky water and finds himself standing in front of a high wall.

With practised skill, he shoots a mini grappling hook from a hand-held gun and scales the wall in seconds.

Before the armed guard patrolling on the other side has had a chance to draw his weapon, he has knocked him off his feet: with a single deft, well-aimed, martial-arts-style kick from his right foot, the guard sprawls motionless on the ground.

The intruder then sprints to one of a series of towering silos, the tops of which are lost in the darkness of the night sky.

He kneels down and runs his hand across the surface until his fingertips sense a concealed button. With a click, a small door pivots open. He scurries inside to a laboratory furnished in the modernist taste. Moving swiftly, he unfastens a three-metre coil of rope from his waist and heads towards a row of large red oil drums menacingly labelled "NITRO".

The rope is not a rope at all, but a long, flexible tube containing plastic explosive that he squeezes, like a lethal toothpaste or the piped icing on a cake, across the oil drums. Next, he swiftly takes off his backpack and opens it to extract an electronic timer. Casting a swift glance at the Submariner strapped outside his diving suit on a webbing strap, he sets the time and primes the device.

Noiselessly, he leaves the mysterious giant silo with its chic Scandinavian-style furniture and drums of "NITRO", escapes detection, scales the wall and is back on the quayside in less than no time.

And then something unexpected takes place.

Rather than returning to the water and making his escape, he peels off his diving suit to reveal a white, single-breasted, wing-lapelled dinner jacket, a crisply laundered, pleated-front dress shirt, and an immaculately fastened, geometrically perfect bow tie. Plucking a fresh carnation that he just happens to have concealed about his person, he puts the flower in the buttonhole of his lapel and walks a short distance into a smoky harbour bar where an exotic dancer moves to the music, seductively inviting patrons to examine her embonpoint.

The man who just minutes ago was wearing a diving suit pulls a slim metal case from his breast pocket, places a cigarette in his mouth and flicks open a Dunhill Rollagas lighter. He surreptitiously permits himself another glance

Released in 1962, *Dr. No* featured Ian Fleming's fictional superspy, James Bond, played by Sean Connery *(pictured opposite with co-star Ursula Andress).*

Following pages: In 1964 Connery starred alongside Honor Blackman in the third outing of the franchise, *Goldfinger*.

at his Submariner by the flame from his lighter.

It is almost twenty past midnight and the sweep-seconds hand with the circular pointer is clearly and easily visible in the gloom of the smoky bar. Just as it reaches the large, inverted triangle at 12 o'clock, a deafening explosion rends the night air and a pillar of yellow flame shoots into the sky.

Roused from their alcoholic torpor, the terrified drinkers stampede for the exit. By contrast, the man in the white dinner jacket with the Submariner is imperturbable. He replaces his lighter in his pocket and exhales a fine plume of cigarette smoke.

It is 17 September 1964, and the audience at the Odeon Leicester Square in London has just seen the first three minutes of what will become the fastest grossing film of its time: *Goldfinger*.

Ten years had passed since the official launch of the Submariner and, having established its credentials underwater, Rolex's Submariner was now imprinting its image on the popular culture of an altogether more glamorous age.

In 1962, Sean Connery had first uttered the now-familiar three-word canticle "Bond ... James Bond", and with it the character from Ian Fleming's espionage novels had been brought to life on the screen. In that film, *Dr. No*, and its 1963 sequel *From Russia With Love*, James Bond had worn a Rolex Submariner – after all, as adverts said, the Royal Navy dived with Rolex, and James Bond 007 holds the rank of Commander in the Royal Navy.

Thanks to the cultural significance that the film franchise would come to assume, the shaken Martini, the Aston Martin and, of course, the Rolex Submariner would become the holy trinity of status-conferring items of the 1960s and beyond, totems to which generations of men would aspire.

And, to prove the truth of the reverse mimetic maxim that life imitates art, by the end of the 1960s, the vocabulary and aesthetic of Rolex advertising was borrowing from the Bond canon. "The Rolex Submariner is a salty watch," explained one advert from the late 1960s. "You'll find it in the cockpits of most ocean racers as hard-driving skippers beat down to Bermuda, Hobart and the Fastnet Rock. How come it's seen so much where the wettest thing is a dry Martini? Who knows. Maybe it's because the black dial goes well with a black tie."[1]

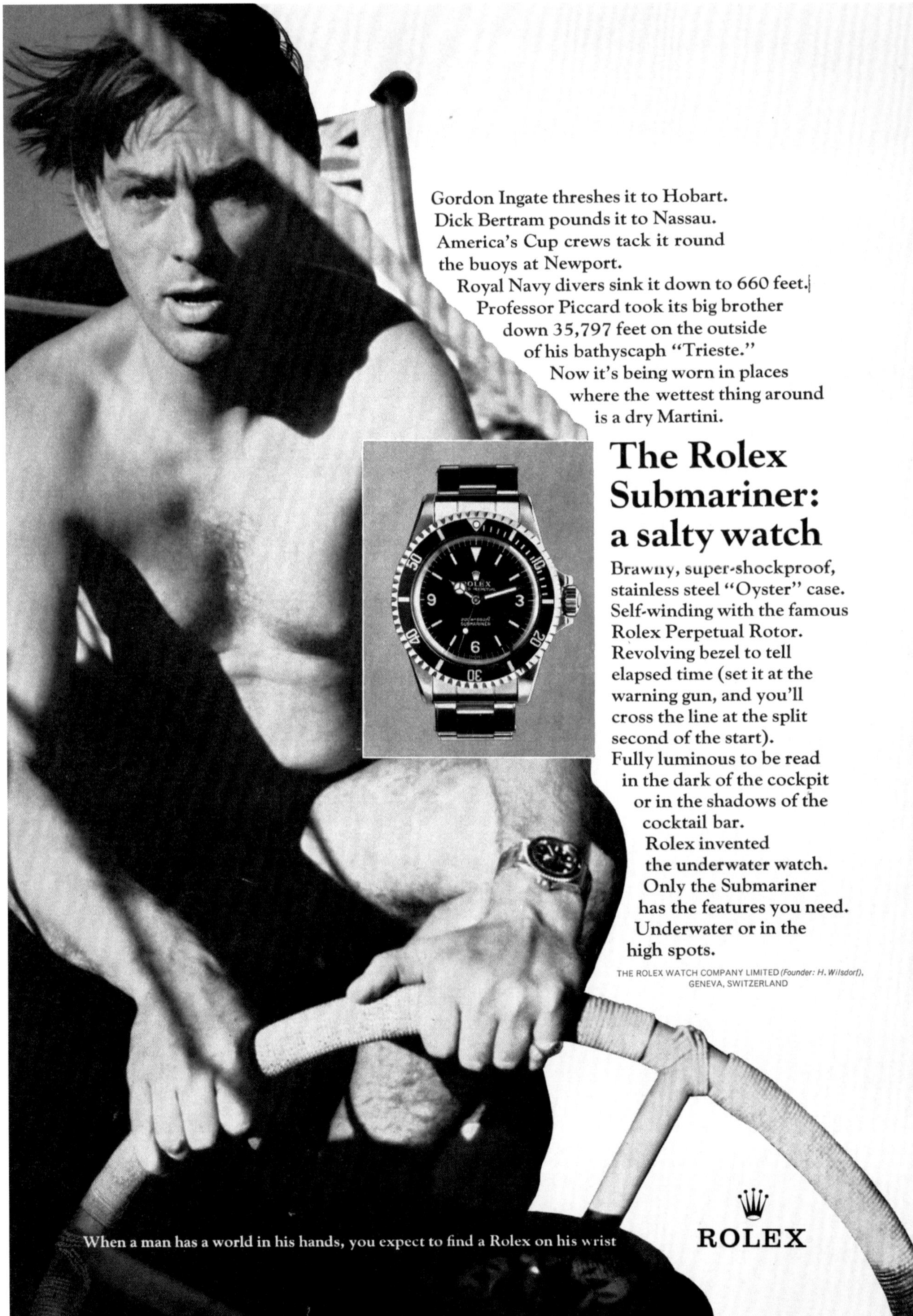
Gordon Ingate threshes it to Hobart.
Dick Bertram pounds it to Nassau.
America's Cup crews tack it round
the buoys at Newport.
Royal Navy divers sink it down to 660 feet.
Professor Piccard took its big brother
down 35,797 feet on the outside
of his bathyscaph "Trieste."
Now it's being worn in places
where the wettest thing around
is a dry Martini.
The Rolex
Submariner:
a salty watch
Brawny, super-shockproof,
stainless steel "Oyster" case.
Self-winding with the famous
Rolex Perpetual Rotor.
Revolving bezel to tell
elapsed time (set it at the
warning gun, and you'll
cross the line at the split
second of the start).
Fully luminous to be read
in the dark of the cockpit
or in the shadows of the
cocktail bar.
Rolex invented
the underwater watch.
Only the Submariner
has the features you need.
Underwater or in the
high spots.
THE ROLEX WATCH COMPANY LIMITED (Founder: H. Wilsdorf),
GENEVA, SWITZERLAND
When a man has a world in his hands, you expect to find a Rolex on his wrist
ROLEX

Above and opposite: A Rolex advertisement from the 1960s borrows from the success of the James Bond franchise by describing the Submariner as a "salty watch" whose popularity extends far beyond its maritime origins.

The decade that had seen the Submariner establish its hegemony over the undersea world had also seen it become an increasingly familiar sight on dry land. In the 1960s, formality and convention were being challenged and a new freedom of expression began to make itself felt in all sorts of ways, even in the way watches were worn. At the beginning of the 1960s, rare was the gentleman who would have even considered wearing a Submariner with a dinner jacket, and yet by the end of the decade, it was most definitely no longer solely considered a professional watch for maritime use. As well as actively marketing it as a watch that wore as well with either a dining or diving suit, Rolex even began to adapt production to reflect the cachet that the timepiece was enjoying on land.

The evolving character of the Submariner was amply demonstrated in a display at the Rolex stand at the 1969 Munich Oceanographic Fair. In a series of vitrines, the development of the Submariner was traced from the first 1926 Oyster through to the 1953 Oyster Perpetual Submariner 100 and the following year's "improved divers' watch, the Rolex Submariner 200" with its Explorer-type dial and "Mercedes" hour hand, the first Submariner equipped with crown guards of 1959, and the *Trieste* "Deep Sea Special" model of 1960.[2]

The next and final cabinet labelled 1969/1970 contained three watches reflecting the growing versatility of the model: the Submariner 200 Date, the helium-escape-valve-equipped Sea-Dweller 2000, and the "18 ct gold Rolex Submariner 200 for divers who want to wear their Submariner always – even for dress occasions."[3]

Agent 007 was arguably the chief agent of the change in perception of the Submariner. James Bond 007 even shares a birthday with the Rolex Submariner. *Casino Royale* was published in April 1953, at the same time that Rolex was in correspondence with Dimitri Rebikoff regarding the development and testing of the Submariner. It was the first in a series of 11 novels and two collections of shorter fiction.

Fleming's literary style was distinguished by its mastery of the material world. He specified the names, brands and manufacturers of everything from extraction fans to cufflinks in order to create a sense of verisimilitude in which to anchor his fanciful tales of espionage and Cold War derring-do.

Although his oeuvre is not celebrated for the Chekhovian depth of its psychological insight, Fleming achieves his

THE ROLEX SUBMARINER REMAINS UMBILICALLY ASSOCIATED WITH THE WORLD'S MOST FAMOUS, LEAST SECRET AGENT.

characterization with an accretion of material detail: through their tastes in food, clothes, cars and accessories, he establishes his characters in his readers' minds, and when it came to his protagonist, he transposed many of his own tastes onto James Bond. The insatiable appetite for Morland cigarettes was one, the wearing of a Rolex was another.

Bond's watch is enough of a personal signifier for Felix Leiter to say: "You're still wearing that old wrist-watch of yours with the big phosphorous numerals."[4] In the 1963 novel *On Her Majesty's Secret Service*, 007 destroys his Rolex while using it as a knuckleduster but he makes himself a promise that he will replace it: "He would get another one as soon as the shops opened after Boxing Day. Another Rolex? Probably. They were on the heavy side, but they worked. And at least you could see the time in the dark with those big phosphorus numerals."[5]

Fleming, who served in wartime intelligence and, like his fictional creation, held the naval rank of Commander, was often photographed wearing a Rolex Explorer Ref. 1016.

In the books, the exact model of Rolex favoured by Bond is never defined, but ever since the first screen outing in 1962, the character has been indissolubly associated with the Submariner, and the sight of the special agent's Submariner filling the screen the second before an explosion became one of the most celebrated – and sometimes parodied – tropes of the wildly successful film series.

Even though the face of the actors playing Bond changed, the face of the Submariner remained a near-constant presence over the course of three decades. The most famous Submariner of all was, as every student of the James Bond canon knows, the bullet-deflecting magnetic version with circular-saw bezel that accompanied Sir Roger Moore on his first interpretation of the role in the 1973 film *Live and Let Die*. It is testimony to the depth of the enduring imprint left upon the collective consciousness by James Bond's Submariners that even though he has not worn one on screen since 1989's *Licence to Kill* starring Timothy Dalton, the Rolex Submariner remains umbilically associated with the world's most famous, least secret agent.

Ian Fleming photographed in his study, 7 November 1963.

Submariner, 1954, Ref. 6538, 38 mm, stainless steel, waterproof to 200 metres (660 feet) worn by James Bond.

Opposite: Agent 007 (first played by Sean Connery, here on the set of *Goldfinger* alongside his Aston Martin DB5) was arguably the chief agent of change in the perception of the Submariner.

THE EXACT MODEL OF ROLEX FAVOURED BY BOND IS NEVER DEFINED, BUT THE CHARACTER HAS BEEN INDISSOLUBLY ASSOCIATED WITH THE SUBMARINER.

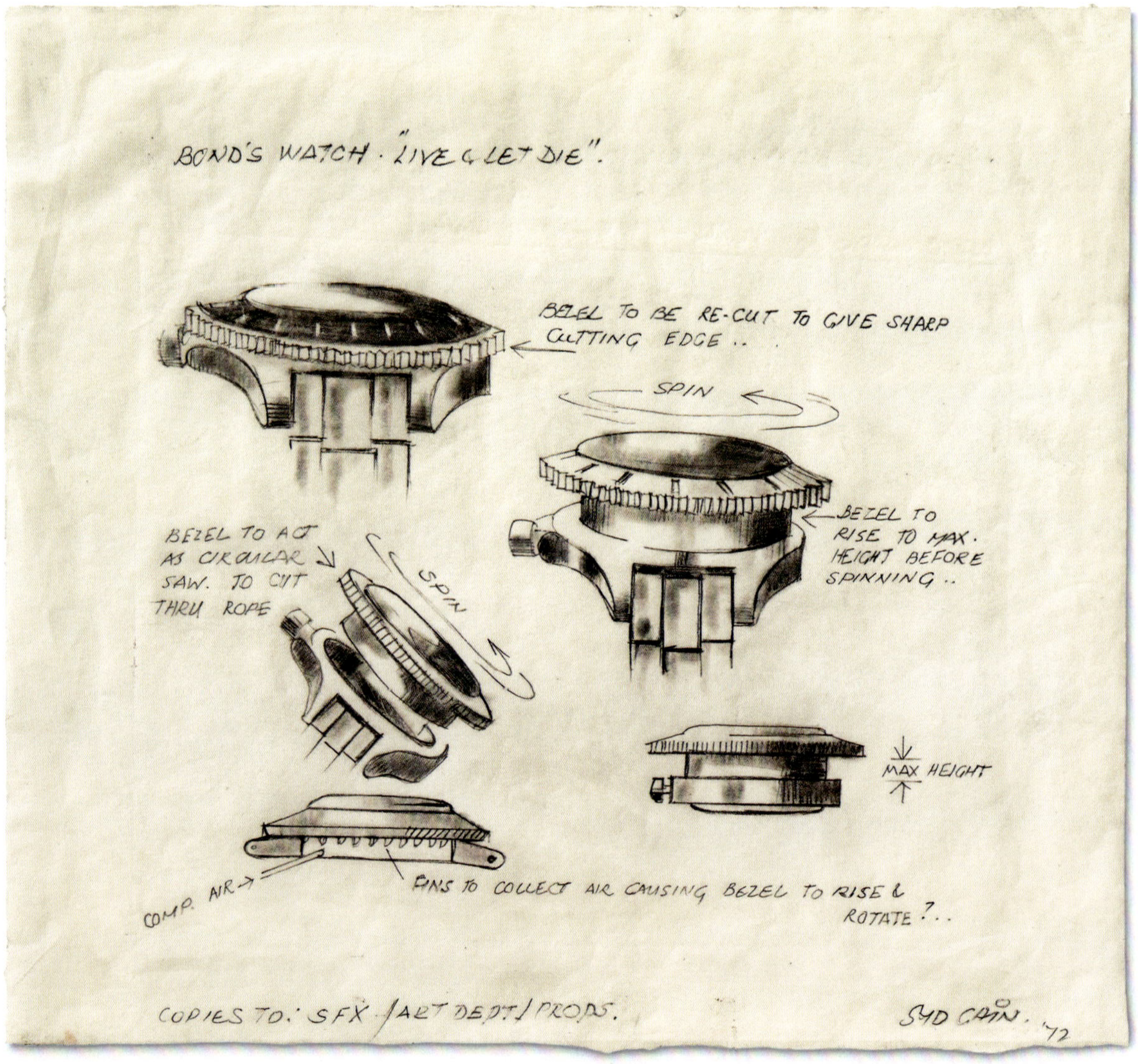

A sketch made by *Live and Let Die* production designer Syd Cain in 1972 shows the transformation of the Submariner's rotating bezel into a circular saw capable of cutting rope for a memorable scene in the film *(opposite)*.

Submariner, 1973, Ref. 5513, 40 mm, stainless steel, waterproof to 200 metres (660 feet). It was on Roger Moore's wrist in the James Bond movie *Live and Let Die*.

660ft = 200m
SUBMARINER

In 1973, Roger Moore assumed the role of James Bond for *Live and Let Die*, in which he starred alongside Jane Seymour.

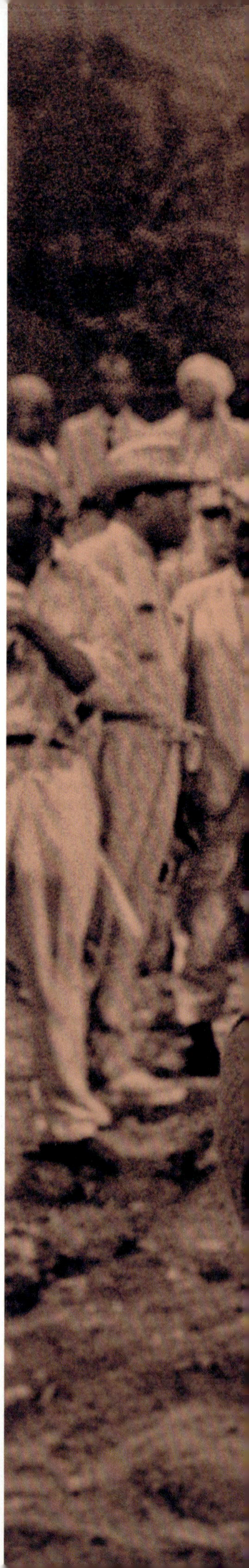

A poster for *The Silent Enemy* (1958), which was filmed off the coast of Malta. It starred Laurence Harvey, who portrayed a celebrated Royal Navy bomb disposal expert who depends on his Rolex watch.

THE SUBMARINER ON SCREEN

Its long-running supporting actor role in James Bond films aside, the Submariner enjoyed a rich and varied career in cinema; a career which began shortly after its launch.

In June 1957, Rolex received a letter from the underwater cameraman on *The Silent Enemy*, starring Laurence Harvey and filming on location in the seas around Malta. Although by no means as famous as the James Bond adoption, the use of the Submariner in this film about the life of a celebrated Royal Navy diver has an important place in the history of the model.

First, the screen appearance takes place shortly after the launch of the watch, and, significantly, the request is made by the underwater filming team, who as well as putting the watch on the star's wrist, were keen to use the timepiece as part of their professional equipment, testifying to what would today be called the "buzz" around this new and exciting type of watch. And then there was the goal of verisimilitude, and as such, *The Silent Enemy* evinced the blurring of the line between fact and fiction, life and art, that would be such a feature of the relationship between Fleming and Bond. However, with this film, the Rolex references were far more complex, not resembling a sequence of events, but rather an intricate historical web.

Laurence Harvey was a rising young actor famous for the air of debonair cynicism and mannered ennui that he conveyed on screen – he would shortly be nominated for an Academy Award for his performance in the 1959 film *Room at the Top*. Produced by Romulus Films, who had Harvey under contract, *The Silent Enemy* was a more straightforward heroic, if somewhat fictionalized, biopic of Royal Navy bomb-disposal expert and diver Lionel "Buster" Crabb.

A hard-drinking, chain-smoking maverick with a taste for loud tweed suits, an aversion to exercise and a penchant for carrying a swordstick topped with a silver crab, he was an unlikely war hero. Crabb had only learnt to dive so that he could defuse underwater explosive devices.

But in 1957, Crabb's name had a new notoriety. One year earlier, he had disappeared while diving in Portsmouth Harbour to spy on a Soviet cruiser that was on a diplomatic mission to the UK. It was one of the more celebrated flashpoints of the Cold War, and at the same time, as Rolex and *The Silent Enemy*'s producers were in correspondence, the story took a further ghoulish twist when on 9 June 1957, a headless, handless body widely believed to be Crabb's was pulled from the water by two fishermen. The state of the remains and the covert nature of Crabb's mission, undertaken at the behest of British Military Intelligence, gave rise to numerous conspiracy theories.*

In a further link to Rolex, in the 1950s, Crabb was appointed Head of the Experimental Clearance Diving Team at the Underwater Countermeasures

* One theory contended that the body was not Crabb's, and that he had been taken to the USSR, brainwashed and set to work training Soviet military divers, a story that has echoes of the plot of another Harvey film, *The Manchurian Candidate*, in which he portrays an American politician who is brainwashed to become an assassin.

Laurence Harvey,
Romulus Films Ltd.,
London. England.

September 30th. 1957.

Dear Mr. Jeanneret,

We have now completed the underwater sequences of my latest film "THE SILENT ENEMY" which, as you know is the story of the wartime exploits of Commander Crabb, the famous frogman.

During the whole of the twelve weeks underwater shooting at Malta and Gibraltar, I , the director William Fairchild and the underwater camera crew have all worn the ROLEX SUBMARINER wrist watch, a watch much favoured by Crabb himself and one which we were advised to use by the expert divers of the Royal Navy who worked with us in the film.

I have nothing but praise for this watch. It kept perfect time all the time, it suffered a great deal from knocks during the underwater fight sequences without any damage at all, it was often subjected to a very high pressure below the surface and even survived quite close explosions from depth charges and other explosives.

During the time we were filming,your product was subjected to every type of stress and strain. Now at home it is still going, quietly and accurately without a scratch on it.

Please accept the thanks of myself and colleagues for the great service your watches did for us personally and for the film as a whole.

Yours sincerely,
Laurence Harvey
LAURENCE HARVEY.

R.P. Jeanneret Esq.,
MONTRES ROLEX S.A.
GENEVE,
Switzerland.

Laurence Harvey

who wears a

ROLEX SUBMARINER

in the film

"The Silent Enemy"

says ***"I've nothing but praise for this watch"***

THE ROLEX SUBMARINER
The watch officially approved by the Admiralty for Royal Naval Divers, is also used by the South African Navy.

SEE THE FILM OPENING TODAY AT THE COLOSSEUM
. . . SEE THE ROLEX WATCHES AT YOUR LOCAL ROLEX JEWELLERS

R 5538

Final Proof Job No. 5538 (JWT) STAR 6 x 5 col. 1958

Produced by Romulus Films shortly after the launch of the Submariner, *The Silent Enemy* was the fictionalized account of Royal Navy bomb disposal expert and diver Lionel "Buster" Crabb. A letter from Laurence Harvey conveys the satisfaction felt by the actor chosen to depict the wartime frogman, pictured on location in Malta.

Opposite: The first Submariner Date, 1969, Ref. 1680, 40 mm, stainless steel, waterproof to 200 metres (660 feet).

Kodak E100 0101
ROLEX
SUBMARINER
SUPERLATIVE CHRONOMETER
28

“IF PETER BENCHLEY CARES TO GO DEEPER STILL IN SEARCH OF HIS NEXT ADVENTURE, HIS ROLEX SUBMARINER WILL NOT BE TROUBLED.”

and Weapons Establishment, which was commanded by Lieutenant Commander Gutteridge, who had worked closely with Rolex on the specifications of the Royal Navy divers' watch.

A film about a much-decorated war hero, diving pioneer, maverick officer in the Royal Navy: the opportunity could hardly have been more propitious for Rolex as the enthusiastic response of 11 June demonstrates:

"The film you are about to make could indeed afford the marvellous opportunity to discreetly, yet dramatically play up the Rolex Submariner, a proven quality product, fully worthy of such a spotlight. This is particularly true in the case of *The Silent Enemy* inasmuch we have very good reasons to believe that when he disappeared, the highly regretted Commander Crabb was wearing a Rolex Submariner."[1]

Richard Winter, Rolex's man in the UK, was also consulted, and his letter throws further interesting light on the close relations enjoyed by Rolex and the Admiralty. "Laurence Harvey and three of his actor-colleagues will be wearing Rolex Oyster Perpetual 'Submariners', which have been offered to them in Malta by the Admiralty and in this connection there may be a possible snag; the Admiralty may be rather loath to part with 'Submariners' and may offer the team 'Explorers'.

"If this is the case, I have arranged for the Art Director in Malta to send me a cable requesting 'Submariners' for the actors (including Laurence Harvey), which I shall pass on to you in Geneva for immediate attention."[2]

Winter also proposed that the production company give Rolex a credit in the opening title sequence. "Quite apart from supplying watches, I have given other assistance to the Assistant Art Director in connection with Commander Crabb himself, whose body, as you will probably have heard, is almost certainly the one that was picked up in Chichester Harbour the other day."[3] According to Winter, "Crabb was wearing an 'Explorer', which I gave him, but unfortunately his body had been in the water for over 12 months with the result that it was headless and handless when found; therefore, the watch has dropped off his wrist and is somewhere in the English Channel now."[4]

It was eventually agreed that photographs of the watches being worn during production by technicians and thespians alike would be supplied to Rolex, "together with a letter from Mr Harvey praising the watches' performances [sic] during the making of the film."[5]

Harvey's letter, more of a panegyric really, could hardly have been more laudatory had Jeanneret written it himself:

Dear Mr Jeanneret,

We have now completed the underwater sequences of my latest film THE SILENT ENEMY, which, as you know, is the story of the wartime exploits of Commander Crabb, the famous frogman.

During the whole of the 12 weeks underwater shooting at Malta and Gibraltar, I, the director William Fairchild and the underwater camera crew have all worn the ROLEX SUBMARINER wrist watch, a watch much favoured by Crabb himself and one which we were advised to use by the expert divers of the Royal Navy who worked with us in the film.

I have nothing but praise for this watch. It kept perfect time all the time, it suffered a great deal from knocks during the underwater fight sequences without any damage at all, it was often subjected to a very high pressure below the surface and even survived quite close explosions from depth charges and other explosives.

During the time we were filming, your product was subjected to every type of stress and strain. Now at home it is still going, quietly and accurately without a scratch on it.

Please accept the thanks of myself and colleagues for the great service your watches did for us personally and for the film as a whole.[6]

The passage of time has decreed that *The Silent Enemy* is remembered as a cinematic curiosity rather than a classic. Yet its use of the Rolex Submariner established the diving watch as more than just a prop: it was a real-world, technical object, the presence of which assisted in the suspension of disbelief. Long before the flash of a Rolex on an actor's wrist became a piece of on-screen semaphore signalling affluence and success, the Rolex Submariner was used to convey a sense of gritty maritime reality, and the most celebrated and enduring example of this triangular relationship between Rolex, cinema and the ocean was to begin in the summer of 1975.

The New York Times described it tepidly as "nonsense, but it's the sort of nonsense that can be a good deal of fun if you like to have the wits scared out of you at irregular intervals."[7] However, at a distance of half a century, it can be seen that the legacy of *Jaws* has proven rather more enduring than the initial response of *The New York Times* would suggest. It was the film that shot Steven Spielberg to fame when he was only 26 years old. Its Oscar-winning combination of bass, cello, trombone and tuba became an aural harbinger of predatory doom. Moreover, it changed the way films were made, marketed and watched. "*Jaws* became the first 'summer blockbuster', redefined how films would be released thereafter, and established a North American distribution and marketing pattern that remains the model for the industry to the present day,"[8] observed screenwriter Carl Gottlieb many years later.

Based on the eponymous best-selling 1974 novel, the story of a giant, man-eating shark terrorizing the waters off an American seaside resort has entered the cultural lexicon much in the way that its author, former White House speechwriter Peter Benchley, has become an enduring part of the story of the Submariner.

His fame as a novelist further magnified by the astonishing success of the film, Benchley featured in advertising for the watch. The most memorable image is the slightly kitsch one of the Submariner dial replaced by a photograph of the bespectacled author, shark in the background, peering out

Advertisement for the Submariner featuring *Jaws* author Peter Benchley, 1975.

Following pages: Actors Jacqueline Bisset, Robert Shaw and Nick Nolte in a scene from Peter Yates's *The Deep*, filmed in Australia in 1977.

from the familiar calibrated black bezel. With the strapline "In the Presence of the Great White ... time suspends itself,"[9] he was identified as the wearer of a gold Submariner.

Another advertisement features the gold Submariner below a picture of a shark, raising, for an instant, the surreal possibility that the shark is the wearer of the watch.

"Peter Benchley always wears his Rolex when he goes fishing"[10] runs the tongue-in-cheek strapline.

"You could say that Peter Benchley invented the great white shark.

"His best-selling book *Jaws* exhibits a near-psychic understanding of the creature's character.

"Even so, when we went down off Dangerous Reef, near Port Lincoln, Australia, to meet the great white shark face to face, it was as heart-stopping and awe-inspiring a moment as any experienced by the characters in his novel.

"Yet while in an underwater cage, under attack from a great white, Peter must have appreciated the supremely functional qualities of his watch; qualities which have made the Submariner standard equipment for divers whose lives depend on split-second timing.

"And if, like Peter Benchley, you ever feel the urge to take a closer look at a shark, the Rolex Submariner has another useful quality.

"It is virtually unaffected by shock."[11]

Jaws was, excuse the pun, such a monster hit that another Benchley book was soon on the bestseller lists and in film production.

The Deep features a slightly more involved narrative than the shark hunt, including shipwrecks, drug smugglers, sunken treasure, a fortune in morphine and a giant moray eel. Production of the film adaptation starring Robert Shaw, Jacqueline Bisset and Nick Nolte took place in 1976, and the finished film offers frequent glimpses of a steel Submariner Ref. 1680 on Nolte's wrist.*

In the months before its release on 18 June 1977, *The Deep* was inescapable thanks to a US$5,000,000 marketing campaign that targeted everything from boat shows to department store cosmetic counters as detailed in an extensive *New York Times* article with the headline "The Selling of *The Deep*",[12] which reported "that every potential moviegoer had read, seen or heard about *The Deep*

* Catalogued as "Property of a Propmaster", Nolte's Submariner Case No. 4,163,403, with the inscription "The Deep 1976" engraved on the case back was sold at Sotheby's Geneva on 11 May 2021.

RSA
ERMUDA

Two decades after *The Silent Enemy* introduced filmgoers to the Submariner, Rolex promoted its central role in the 1977 seaborn thriller *The Deep*, based on the novel written by Peter Benchley.

at least 15 times"[13] before buying a cinema ticket. There was even an "original rum drink to be called 'The Deep'"[14] intended for consumption "especially in singles' bars."[15]

As *The New York Times* made clear, the Rolex Submariner was in the eye of this global publicity storm. "Then there was the Rolex Sub-Mariner [sic] watch, utilized underwater by the actors to measure minutes of oxygen remaining in tanks. Rolex distributed 250,000 flyers to its thousands of retailers. There were TV commercials showing the stars wearing the watch."[16]

The film's title was of course a gift for advertising copywriters.

"Peter Benchley's Rolex is guaranteed against *The Deep*.

"When Peter Benchley's *Jaws* was filmed, it broke all box-office records as the most successful motion picture in history.

"Now the film of Benchley's latest best-selling novel *The Deep* has set new standards for realism and excitement in underwater action photography.

"Shot on location around the coral-encrusted wreck of the Royal Mail ship *Rhone* off the British Virgin Islands and off Bermuda, nearly half the film takes place beneath the ocean's surface.

"Altogether, cast and crew spent 10,780 man-hours underwater. And like Peter Benchley, the watch every one of them wore was a Rolex Submariner.

"Regarded as standard equipment by divers throughout the world, the Rolex Submariner is a superb example of helvetic craftsmanship. It is an Officially Certified Swiss Chronometer, indispensable for people whose lives depend on perfect timing.

"The Oyster case and crystal are completely proof against water; further pressure merely increases the efficiency of the seal. And the Twinlock winding crown actually screws down on to the case like a submarine hatch.

"Much of the action in *The Deep* happens 30 metres below the surface.

"But if Peter Benchley cares to go deeper still in search of his next adventure, his Rolex Submariner will not be troubled.

"It is guaranteed to a depth of 200 metres."[17]

As with *The Silent Enemy* two decades earlier, the Submariner had been used by cast and crew alike. Given the success of *Jaws*, Rolex was ready to exploit the film to the full. In addition to the new advert, there were promotional activities across the world: the files in the Rolex Archives burst with press clippings, articles, memos, minutes of meetings, and letters relating to the role of Rolex in *The Deep*.

A special Rolex booklet concerning the Submariner and *The Deep* was produced to coincide with the release in cinemas, as was a 30-second commercial showing the Submariner in the film. From America to Japan, where a sports stadium in Tokyo hosted a mass screening, it seemed that no market for Rolex escaped the exhaustive publicity blitz. The brand's then-advertising agency wrote to Rolex offices to make them aware of this blizzard of activity.

"The success of promotional tie-ups between Rolex and Columbia Pictures in the US and Japan shows how valuable this activity can be in your territory.

"Your local Columbia Pictures Publicity Manager will be contacting you about exploiting the link between Rolex and *The Deep*."[18]

Indeed, Columbia's PR team were tireless in their efforts on behalf of this film.

"Your Columbia Pictures representative has been instructed to work very closely with you. There are endless opportunities for organising promotions in your mutual interest, and Columbia Pictures can supply you with material, like copies of the book, colour and black-and-white stills, and so on. Where possible, opportunities may exist for him to help you arrange special screenings of *The Deep* for Rolex jewellers and customers."

Although the metaphor may not be entirely appropriate for a film that mainly takes place underwater, as far as Columbia Pictures was concerned, the sky was the limit as regarded publicity.

"*The Deep* is big. Very big. So the more promotional tie-ups that can be arranged with Columbia, the better."[19]

All around the world, retailers acted on these instructions.

"Enclosed please find nine pages from 'Ekstra Bladet', at [sic] which you will find 'Rolex Submariner' described" ran a typical letter from a Rolex retailer in Copenhagen.

"This competition has without doubt been a very good advertising for Nick Nolte and Rolex. We now will follow up with a special exhibition in our show window at the entrance to the movie, where *The Deep* is being shown.

"Besides we intend to run a couple of advertisements the same day as the film starts."[20]

Like so many others, this Danish retailer was well aware of and impressed by the power of the unique relationship that the Rolex Submariner enjoyed with the world of cinema.

THE SUBMARINER ENJOYED A RICH AND VARIED CAREER IN FILM; A CAREER WHICH BEGAN SHORTLY AFTER ITS LAUNCH.

Set in London in the Swinging Sixties and influenced by the lives of model and photography legends of the time, Michelangelo Antonioni's *Blow Up* (1966), starring Vanessa Redgrave with a Submariner on her wrist, has become a cult classic.

In *All the President's Men* (1976), Robert Redford plays *Washington Post* reporter Bob Woodward alongside Dustin Hoffman as Carl Bernstein. The two journalists became known for breaking the White House Watergate scandal and Nixon's subsequent impeachment. Redford wore his own Submariner in the film.

In *The Punisher* (2004), John Travolta plays a crime boss. He is featured wearing a yellow-gold Submariner.

Luke Wilson wears a yellow-gold Submariner, Ref. 116618 LN, with a black dial in *Middle Men* (2009).

In the action comedy *Knight and Day* (2010), Cameron Diaz wears a Submariner, Ref. 16610.

Jim Carrey is cast as a real estate entrepreneur who inherits a penguin in the 2011 comedy, *Mr. Popper's Penguins*. He wears a yellow Rolesor version of the Submariner.

Based on the book by Michael Lewis, *The Big Short* (2015) became a cult classic as a drama/comedy portraying the events leading up to the global financial crisis of 2007–2008. In it, lead actor Steve Carell wears a Submariner watch.

American actor Steve McQueen had a great fondness for wearing a Submariner in life and in film. He played Michael O'Halloran in the celebrated disaster movie *The Towering Inferno* (1974).

CHAPTER XII

AN IMMUTABLE ICON

The Deep was directed by Peter Yates, who had become famous almost a decade earlier as the director of the 1968 hit *Bullitt*, which had also starred Jacqueline Bisset, this time alongside Steve McQueen. McQueen's name is, of course, most commonly associated with the Explorer II, famous for its orange 24-hour hand. However, evidence to support this connection is scarce to the point of non-existence, whereas his fondness for the Submariner (without a date) is attested to by frequent appearances on his wrist both in film and in private life.

The Submariner in question could be the Ref. 5512, and it can be seen on the actor's right wrist in the 1974 disaster film *Towering Inferno*, in which he portrays a fire chief, and again in his last film, *The Hunter*, released in 1980, in which he is cast as an ageing bounty hunter.

Given the enduring status that McQueen enjoys as "king of cool" and prototypical, pre-gender-equality icon of unreconstructed masculinity, the appeal that the robust, unadorned functionality of the Submariner had for him is unsurprising.

More noteworthy is that the Ref. 5512 was introduced in 1959, followed by the almost identical Ref. 5513 in 1962 that last appeared in the 1989/1990 catalogue – a record in longevity. To all but the most minute inspection, the same watch that had been launched in the late 1950s was still being made on the threshold of the century's final decade.

It is a truism to observe that the world underwent significant change between 1958 and 1989. In 1958, the world witnessed the first transatlantic jet flights, Nikita Khrushchev became leader of the USSR, and Dwight D. Eisenhower was in his second term as President of the U.S. The year 1989 was the year that George H. W. Bush became President of the U.S., the year that the Berlin Wall fell, effectively ending the Cold War, and the year that Sir Tim Berners-Lee invented the World Wide Web.

But, as the world changed beyond recognition, the Rolex Submariner remained almost immutable. The development and early years of the Submariner had seen changes on an almost monthly basis, as input arrived from sources as diverse as British navy divers in cold Scottish lochs and eccentric inventors in the warm Mediterranean. Between 1953 and 1958, seven different references of Submariner were put into commercial production*, and prototypes circulated amongst military, scientific and civilian divers.

During this period, the case diameter was 38 mm, but other specifications were subject to change: crown diameter fluctuated between 5.3 and 8 mm; dial design varied, sometimes using Arabic numerals and featuring varying amounts of text and colour of print; hand size and shape evolved; bezel calibration likewise altered between models.

But as the 1950s drew to a close, the frantic pace of innovation slowed and the familiar lineaments of the Submariner emerged in the form of the Ref. 5512: the 200 metre (660 feet) water resistance; the rotating bezel that functioned using a press-and-turn system rather than relying purely on friction; the black aluminium bezel insert with the inverted triangle and spot of luminous material followed by minute indicators until 15, and at five-minute intervals thereafter; the crown guards that swelled from the straight-sided case wall to embrace the seven-millimetre diameter winding crown; the robust 40 mm

* Ref. 6204 (1953), Ref. 6200 (1954), Ref. 6205 (1954), Ref. 6538 (1954), Ref. 6536 (1955), Ref. 5508 (1958), Ref. 5510 (1958).

GIVEN THE ENDURING STATUS THAT MCQUEEN ENJOYS AS “KING OF COOL” AND PROTOTYPICAL, PRE-GENDER-EQUALITY ICON OF UNRECONSTRUCTED MASCULINITY, THE APPEAL THAT THE ROBUST, UNADORNED FUNCTIONALITY OF THE SUBMARINER HAD FOR HIM IS UNSURPRISING.

Actor Steve McQueen wearing a Submariner whilst competing in the 12 Hours of Sebring endurance race, 21 March 1970.

A FEAT OF HOROLOGICAL ENGINEERING, THE SUBMARINER SAYS SOMETHING ABOUT ROLEX'S APPROACH TO WATCHMAKING AND, BEYOND THAT, THE WORLD AT LARGE.

case; the Mercedes hour hand and sweep-seconds hand with the circle of luminous material near its end ... these were features that would remain constant until well into the following century and come to define and enshrine the Submariner in the popular imagination.

The simple, empirically established design gained its cultural power through its constancy and repetition hundreds of thousands of times as, year after year, Submariners entered the market and, thanks to their solidity and near indestructibility, remained in use and in circulation over years then decades and then generations, cumulatively reinforcing the archetype with every watch produced.

The unchanging strength of the design was further enhanced and thrown into relief by the horological context in which it was produced. The Ref. 5513 spanned an incredibly turbulent period in watchmaking. During this time the industry underwent radical and enduring change. In 1962, when it appeared, Swiss watchmaking was entering a boom period as the near-unchallenged master of the industry. But at the beginning of the 1970s, a concatenation of circumstances, including the arrival of new technology, the rise of inexpensive, hyper-accurate, quartz-regulated electronic wristwatches, and the soaring value of the Swiss franc, almost destroyed Swiss watchmaking. Brands, many older than Rolex, simply went out of business. Of those that survived, many changed their structures and altered the designs of their key products with bewildering rapidity to take account of swift changes in technology, fashion and design in the hope of appealing to new markets and new customers. By the time that the Ref. 5513 was discontinued at the end of the 1980s, it is only a slight exaggeration to say about the only aspect of watchmaking that had remained constant and unchanged to the millimetre, if not the micron, during that period of immense upheaval was the design of the Rolex Submariner.

That is not to say that the Submariner did not evolve during those decades. As has been seen, 1969 welcomed the gold Submariner and the appearance of the date function in the Ref. 1680, enhanced by the signature addition of the Rolex Cyclops lens. The appearance of a luxury model equipped with a function utterly irrelevant to the practical needs of the diver testified to the growing popularity of the Submariner as a mainstream model for daily wear on dry land.

Submariner, 1959, Ref. 5512, 40 mm, stainless steel, waterproof to 200 metres (660 feet). The model now features an integral crown guard.

If it's good enough for a submarine,

This is the Triplock winding crown of a Rolex Submariner. It screws down on the solid Oyster case to close as securely as the hatch of a submarine, employing three separate seals in

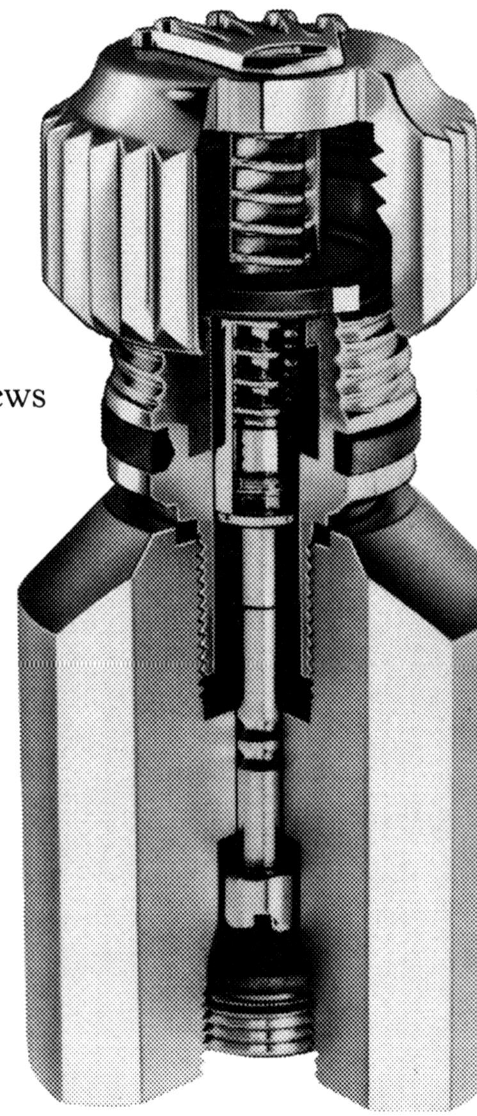

the process. The result is a watch that is watertight to 1,000 feet. A depth that very few divers – and, for that matter, few submarines – are ever likely to encounter.

it's good enough for a Submariner.

1953 • 100 M

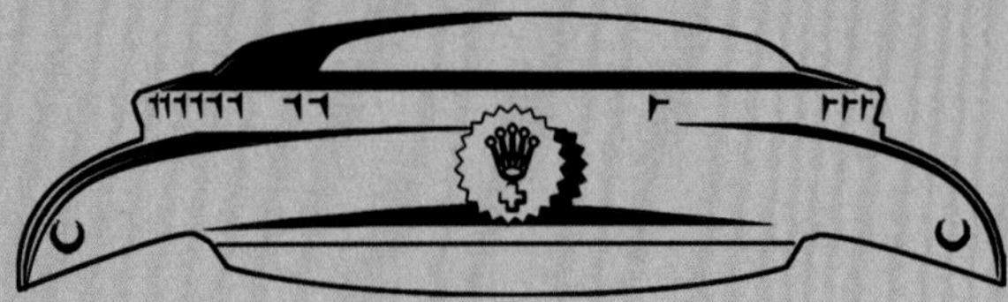

REF. 6204
Launch of
the Submariner

1954 • 200 M

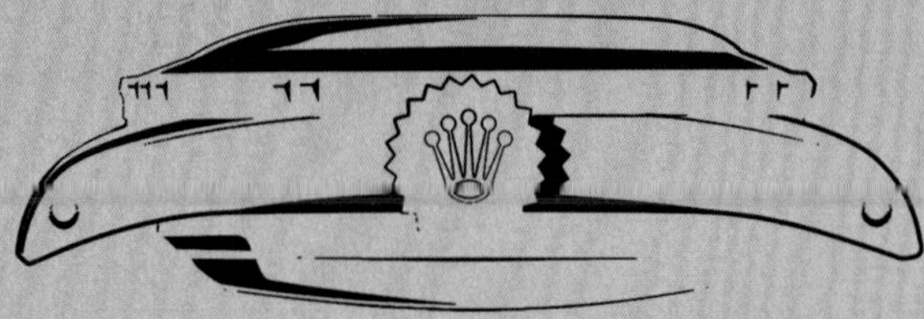

REF. 6538
Winding crown, 8 mm

1959 • 200 M

REF. 5512
Integral crown guard

1969 • 200 M

REF. 1680
Cyclops lens

1979 • 300 M

REF. 16800/16808
Scratch-resistant
sapphire crystal

2003 • 300 M

REF. 16610 LV
Green bezel

2008 • 300 M

REF. 116618/116619 LN/LB
Cerachrom bezel

2020 • 300 M

REF. 1266XX
41 mm, crystal with
anti-reflective coating

Above and opposite: From 1953 to 2020, and beyond. Fluted winding crown and case back, bezel with knurled edges, contrasted numerals and graduations, hour markers in the shape of triangles, circles and rectangles, broad hands: these are all timeless features that ensure legibility and reliability underwater. The Submariner has constantly improved over the years while remaining faithful to the original aesthetics. The evolution continues.

Calibre 3135 that equipped the Submariner Date until 2019.
Opposite: Submariner Date, 1986, Ref. 16610, 40 mm, 904L steel, waterproof to 300 metres (1,000 feet).

In 1970, the Triplock crown made its debut in a commercially produced model. Its predecessor, the Twinlock winding crown introduced in 1953, had represented a significant improvement on the 1926 Oyster. As the name suggested, it offered a double waterproofness system with two sealed zones: one inside the crown tube, the other inside the crown. It also brought the ancillary benefit of providing further support for the winding stem, enhancing the overall reliability and longevity of the watch. The Twinlock system continues to be used on many non-Professional Rolex models to this day.

The Triplock crown was first used on the Sea-Dweller because of the unique pressures to which the deep-diving model would be subjected. Its chief feature was an additional O-ring inside the tube, effectively creating an airlock that improved the integrity of the case. The Triplock crown was first fitted to the Submariner in 1977 and would in time appear in a number of other Professional models.

Two years later, in 1979, the operational depth of the Submariner was increased to 300 metres (1,000 feet) notably with the introduction of sapphire glass. The same year saw the implementation of a new unidirectional bezel system that allowed moving the bezel in definable increments, in order to increase the diver's safety. Each slight movement generating the satisfying click that would become such a familiar feature for millions of Submariner wearers.

From time to time, the lineup was enlivened by new materials. Rolesor, the proprietary name for Rolex's combination of steel and gold, was introduced in the model in 1983, and in 1985, the steel used for Submariner cases was upgraded to 904L, an alloy that contains a higher proportion of chrome, increasing its resistance to corrosion, especially by seawater.

Although the venerable Ref. 5513 eventually bowed out after more than a quarter of a century in production, the various contemporaneous and successor references of Submariner continued to faithfully observe the aesthetic codes with only the most subtle of changes to incorporate new technology and calibres, as, eventually, the trusted 1520 and 1530 calibres were replaced by the 3035 and 3135.

And then, in the early years of the 21st century, the Submariner underwent the most startling visual change in its entire history.

ROLEX
PERPETUAL DATE
SUBMARINER
SUPERLATIVE CHRONOMETER

Left: Submariner Date, 1983, Ref. 16803, 40 mm, yellow Rolesor version, waterproof to 300 metres (1,000 feet).
Right: Submariner Date, 1969, Ref. 1680, 40 mm, 18 ct yellow gold, waterproof to 200 metres (660 feet).

THE SIMPLE, EMPIRICALLY ESTABLISHED DESIGN GAINED ITS CULTURAL POWER THROUGH ITS CONSTANCY.

Right: Design drawing of the Submariner Date Ref. 126610 LN.
Far right: Fitting of the hands on a Submariner Ref. 126610 LN.

Following pages, left: Triplock winding crown of the Oyster Perpetual Submariner.
Centre: Stamping of a Submariner middle case. *Right:* Starting in 2016, all Rolex watches undergo a waterproofness test in hyperbaric tanks.

WATERPROOFNESS TEST

AS PART OF THE SUPERLATIVE CHRONOMETER CERTIFICATION, THE WATERPROOFNESS OF EACH WATCH IS VERIFIED BY IMMERSING IT IN WATER IN A HYPERBARIC TANK DEVELOPED IN PARTNERSHIP WITH COMEX.

A general view of the waterproofness test section of the Superlative Control. When a watch has passed the Superlative Control tests that bring the manufacturing process to an end, it is certified as a Superlative Chronometer.

The Superlative Chronometer certification, unique to Rolex, is symbolized by a green seal and coupled with an international five-year guarantee.

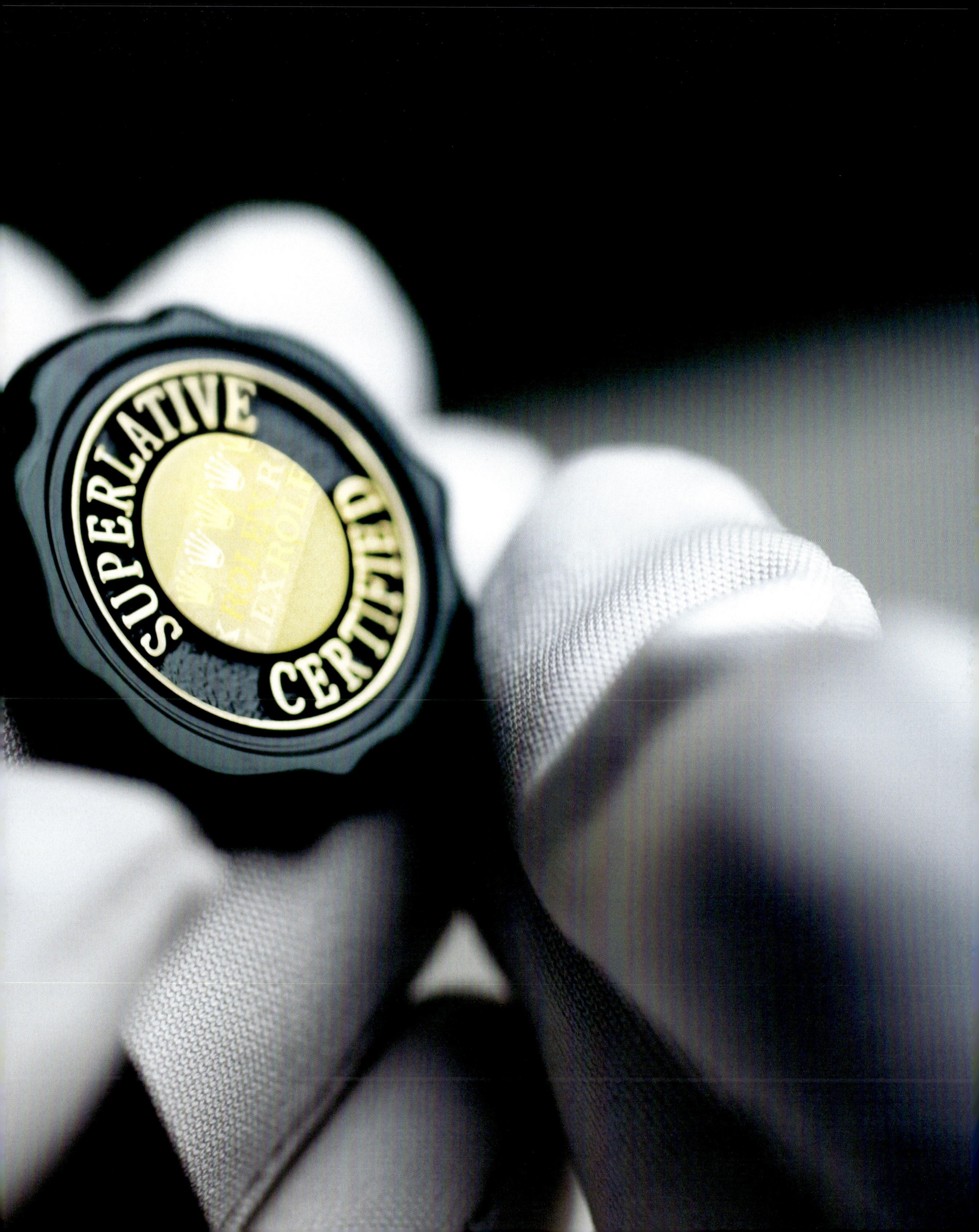
SUPERLATIVE
CERTIFIED

DAVID DOUBILET

Rolex Testimonee

A true inheritor of the legacy of Dimitri Rebikoff and Jacques-Yves Cousteau, David Doubilet has spent half a century recording submarine life in all its rich splendour and dazzling diversity. A child of the aqualung age, Doubilet's first photograph was published in *National Geographic* in 1972. Over the intervening years he has continued the mission of those early pioneer diver-photographers who first returned from the waves bringing tales of a colourful and exotic underwater world and the images to prove it.

Much may have changed since Rebikoff made his first trepidatious dive in the Bay of Cannes, not least the sophistication of the diving and camera equipment at Doubilet's disposal. But essentially the mission is still the same. And today that mission is more urgent than ever. Most people's experience of the world's oceans continues to be mediated by the photographer's eye and the camera's lens. Whereas members of the pioneer generation were awestruck and simply marvelled at the photographic potential of this new realm into which man was only just beginning to set foot, today Doubilet finds himself burdened with responsibilities that are more than artistic: "My work has to begin with the visual possibilities, but beyond that conservation issues are the driver."[1]

He is fascinated by the majesty and fragility of life in the world's oceans, and he is particularly drawn to coral reefs and icebergs. In particular he recalls one poignant return to the Great Barrier Reef where earlier he had recorded an abundance of colour and life: a polychrome

Left: Barracudas circle diver Dinah Halstead in the clear waters of Papua New Guinea, 2009. *Far left:* David Doubilet wearing the Oyster Perpetual Submariner, 2016.

"I HAVE LEARNED THAT IMAGES HAVE THE POWER TO EDUCATE, HONOUR, HUMBLE AND ILLUMINATE."

DAVID DOUBILET

ecosystem bursting with colourful fish cloaking a robust reef. "We returned after sequential coral bleaching events to discover a silent coral cemetery. Even in one of the best protected settings in the world, reefs change constantly. A storm, a rise in water temperature beyond what the corals can tolerate or a shift in sea level – all of these things can change a reef very, very rapidly."[2]

He finds icebergs equally hypnotic. "Icebergs mesmerize me because they are a perfect metaphor for the sea: a small fraction visible to the naked eye. Greenland's iceberg garden at Red Island in Scoresby Sund Fjord is a place where beautiful iceberg sculptures tell a very ugly truth about glacier retreat. I am interested in putting a face on climate change that no one can ignore."[3]

He became a member of the Rolex family in 1994, and since then he has worked with other Rolex Testimonees including photographing the grande dame of the deep, Sylvia Earle.

Whether frolicking sealions in the sun-dappled waters surrounding the Galápagos Islands; the controlled menace of a school of sharks circling an intrepid diver; a sleek bullet-fast penguin diving off an Antarctic iceberg; or the joyous splash of bright orange banded by black and white stripes that is the clownfish: Doubilet's images transport the viewer to the aquatic world where wonder is clouded by worry over the future of this fragile world. Doubilet's photographs are superb and evocative records of the marine environment, but more than simply documenting the world's oceans, Doubilet, an enthusiastic Testimonee for Rolex's Perpetual Planet Initiative, is working hard to ensure that the underwater world that he has been fortunate enough to experience and document will survive to be studied, nurtured and enjoyed by generations as yet unborn.

Underwater photographer David Doubilet, a Rolex Testimonee since 1994.

A diver kneels beneath the bow of Russian destroyer 356, Cayman Brac, Cayman Islands.

The *DEEPSEA CHALLENGER* piloted by James Cameron in 2012 was the first manned submersible to reach the bottom of the Mariana Trench since the *Trieste* in 1960.

CHAPTER XIII

THE PAST TWENTY YEARS

"I would like to see something special for this anniversary."[1]

In January 2003, Rolex CEO Patrick Heiniger called a meeting with Rolex's Creation Department. It was half a century almost to the day since Hans Wilsdorf had written to René-Paul Jeanneret observing that the proposal of a divers' watch "seems to open fresh possibilities for Rolex becoming well known for our waterproof watches".[2]

Even its visionary founder could never have imagined just how well known for its waterproof watches Rolex would become. What is more, this fame had been earned by adhering to the aesthetic codes and industrial parameters he had established. Had the ghost of Hans Wilsdorf floated among the glass-walled towers of the Rolex headquarters in Geneva at the beginning of the 21st century, he would have recognized the Submariner immediately.

It was the first time anyone in the department had been asked to celebrate an anniversary. Moreover, as well as coming as a surprise, the request came with an uncharacteristic sense of urgency. Heiniger wanted to have a watch ready for the annual watch fair in Basel, rebranded Baselworld that year, which opened at the beginning of April. By 2003, Rolex was a sizeable industrial concern manufacturing almost three-quarters of a million watches a year, a level of production that required considerable forward planning. A new watch would take years to launch, and yet here was the CEO breezily asking for an anniversary model. Furthermore, he gave his creative and technical teams little more than two months to have it ready.

The result was the famous Ref. 16610 LV*, identifiable at a distance by its green bezel insert and, on closer inspection, by slightly larger hands and indices. This was a startling departure from the half-century tradition of predominantly black inserts. The very identity of the watch was founded in great part on the monochrome palette and the functional contrast between black and white. Now, after 50 years, here was a Submariner that purported to celebrate that tradition by flouting it.

The model was an instant hit, becoming one of the most discussed and sought-after watches of the era. Such was its success that it became much more than a commemorative watch. Instead of entering production for the anniversary year, the green-bezelled Rolex Submariner would become a part of the main collection.

Although the concept of a commemorative Submariner may have been an idea that had occurred to Heiniger over the 2002 to 2003 Christmas and New Year's holidays, ingeniously interpreted and swiftly executed by the creative and technical teams, it would prove an enduring contribution to the canon.

By the early 21st century, watches and the way they were being worn were changing. Interest in historic models was awakening. The Rolex Submariner found itself in the unusual position of being an icon anchored in a historic period, which, thanks to its uninterrupted production and incremental improvement, was simultaneously very much a part of the contemporary watch industry. It was a strong pillar of both Rolex's production and the wider edifice of Swiss watchmaking. The green bezel model of 2003 was the conceptual resolution of these two aspects of its character.

* LV: Lunette verte (green bezel)

THE GREEN BEZEL WAS INTRODUCED TO MARK THE 50TH ANNIVERSARY OF THE SUBMARINER.

Submariner Date, 2003, Ref. 16610 LV, 40 mm, 904L steel, waterproof to 300 metres (1,000 feet).

ROLEX
SUBMARINER
SWISS MADE

Submariner Date, 2010, Ref. 116610 LV,
40 mm, 904L steel, waterproof to 300 metres
(1,000 feet). For the first time, the model features
a Cerachrom bezel insert in green high-tech ceramic.

Moreover, the precedent established by this watch would have consequences that would reach far into the future and open up new opportunities for the brand.

But, while the green bezel insert was a potent symbol, it was also just a flimsy ring of coloured aluminium – a fact that had not gone unnoticed in Rolex's technical department.

Aluminium bezel inserts had been used for 50 years at Rolex. But as case engineering and crystal quality improved over time, the shortcomings of this material began to become more apparent. Over long periods of sustained use, the aluminium was vulnerable to corrosion, susceptible to colour changes from ultraviolet light, and could be scratched. For a company that prided itself on the structural integrity of its watches, these were obvious shortcomings. And it was to rectify these shortcomings that, five years later, the era of the ceramic-bezel Submariner began.

As is customary, the material was given a Rolex name and christened "Cerachrom" – an elision of ceramic and chromatic. The Submariner range was equipped with a Cerachrom bezel insert in 2008. The green Cerachrom bezel insert was introduced two years later.

The ceramic bezel inserts are made of zirconium dioxide, a material derived from zircon, known for its hardness. The powdered ceramic preparation is mixed with binding agents, heated and injection moulded to create the essential shape along with numerals, scales and inscriptions. Heat treatment removes the binding agents, and during sintering – a 24-hour firing process at up to 1,600 degrees Celsius – the bezel assumes its final colour and shrinks by 25 to 30 per cent. The numerals and scales moulded in the ceramic are coated with a micron-thin layer of gold or platinum using a PVD process. A final diamond-polishing removes the gold or platinum from the rest of the bezel's surface and imparts the long-lasting, highly resistant lustre.

Indeed, the new bezel material was just one of many changes, additions and improvements that the Submariner range would witness in 2008, chief among which was the appearance of the 44 mm, an underwater behemothic dreadnought known as the Rolex Deepsea Ref. 116660. As an ongoing evolution of Rolex's dominance of the world beneath the waves, visually and conceptually, this was classic Rolex. Waterproof to a staggering depth of 3,900 metres (12,795 feet) and tested with a safety margin of 25 per cent according to the standard in effect for diving watches, the Rolex Deepsea demonstrated Rolex's continuing commitment to the battle against its old foe: water. In one way, the Rolex Deepsea was the definition of horological over-engineering: it was designed and built to function under conditions that were unequivocally lethal for human beings. The 100 metres (328 feet) of water resistance afforded by the very earliest Submariner remains sufficient for all but the most extreme use, and yet the Rolex Deepsea offered 39 times as much protection.

The exceptional water resistance with which this watch was endowed was achieved using a patented case architecture that Rolex named the "Ringlock system". Literally and figuratively at the centre of this system is a compression ring made of incredibly resistant nitrogen-alloyed steel, on top of the ring is a thick domed sapphire crystal, and at the opposite end of the ring is a case back of titanium alloy. The Ringlock system functions as an inner case fitted into a familiar, albeit slightly enlarged, Oystersteel case. Rolex had mapped the extreme pressures of several tonnes that would be brought to bear on the watch at its maximum stated operating depth and had determined that at great depth, the majority of the pressure was exerted on the top and bottom of the watch, that is, the titanium back and thick crystal front; pressure that would be borne by the supremely hard compression ring.

But it was not enough that this watch should be able to withstand such conditions; it had to be able to function at such depths, and, of course, the primary duty of any watch is to convey the time to the wearer (even though any human wearer would be long dead at this depth).

What's more, the Rolex Deepsea was equipped with a new invention to enhance its legibility in all circumstances: the Chromalight display. Since their launch, the Submariner models have featured a luminescent display. Rolex had been using Super-LumiNova since 1998, but in 2008, with the launch of the Rolex Deepsea, the Chromalight display was unveiled. Recognizable by its blue glow in the dark, this display boasted a duration of luminescence almost twice that of the standard phosphorescent material of the day. In addition, its intensity during emission was improved.

SEA-DWELLER
OFFICIALLY CERTIFIED
28

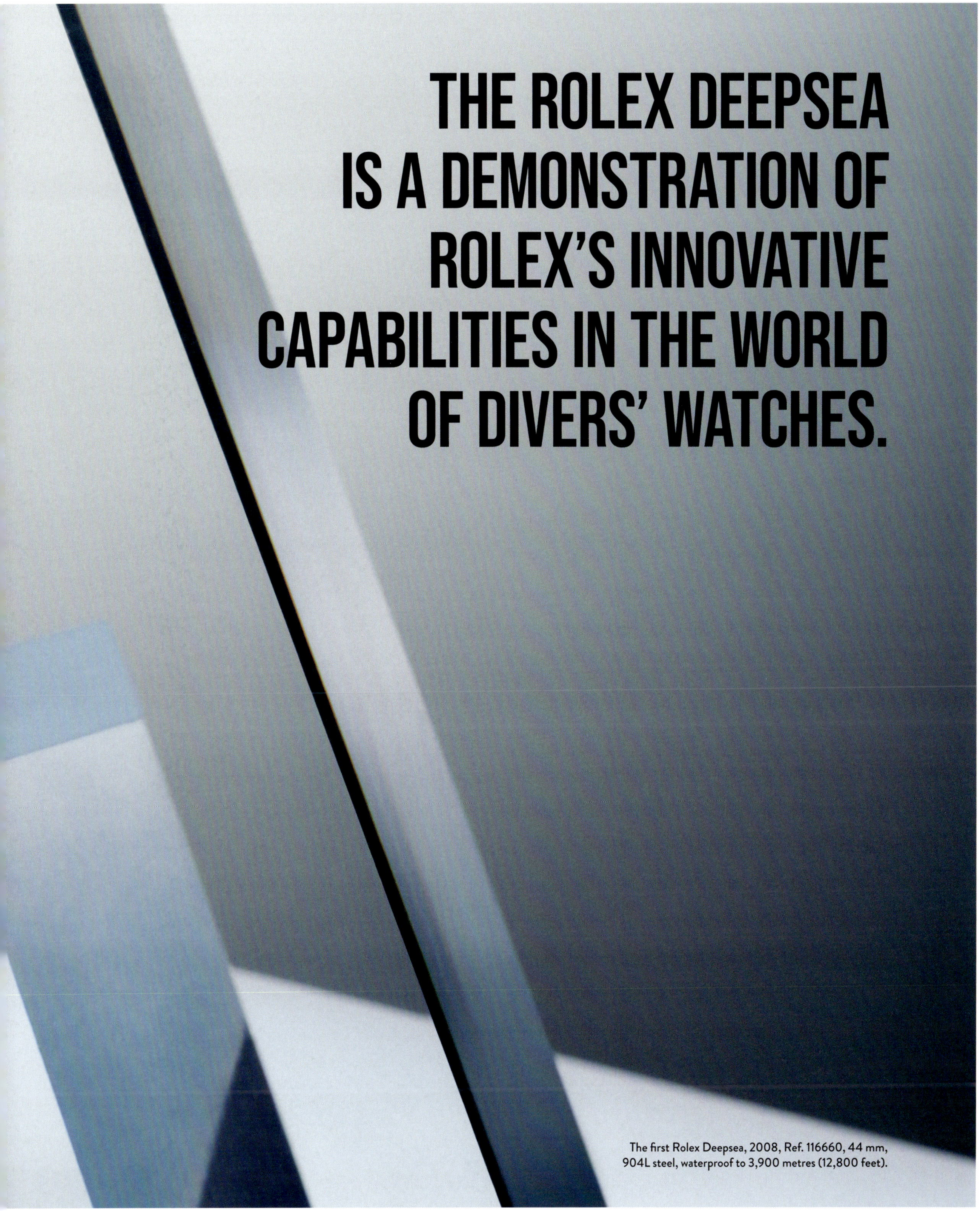

THE ROLEX DEEPSEA IS A DEMONSTRATION OF ROLEX'S INNOVATIVE CAPABILITIES IN THE WORLD OF DIVERS' WATCHES.

The first Rolex Deepsea, 2008, Ref. 116660, 44 mm, 904L steel, waterproof to 3,900 metres (12,800 feet).

"THE ONLY DIVING INSTRUMENTS WHICH PERFORMED ALL THE TIME WERE OUR ROLEX WATCHES."

EMMANUELLE PÉRIÉ-BARDOUT

Moreover, given the extra bulk of wetsuits and the effect on the wrist of pressure at the variety of depths at which the Rolex Deepsea could be used, a new, rapid system for changing the length of the bracelet was introduced: its name was "Rolex Glidelock".

Two years later, the Rolex Deepsea was able to demonstrate its capabilities when, a century after the conquest of the North Pole, eight "pioneers of extreme frontiers" undertook a 45-day polar expedition, Deepsea Under The Pole by Rolex, that was dedicated to furthering knowledge of the ecosystem of the waters beneath the Arctic sea ice. Although these eight explorers would not, of course, descend to the maximum depth of the Rolex Deepsea, the watches with which they were issued were subjected to stern trials both in and out of the water as their intrepid wearers ski-trekked across the white wilderness at the top of the world, stopping periodically to make dive after punishing dive.

Speaking on her return, Emmanuelle Périé-Bardout, the team's only female member, said: "The only diving instruments which performed all the time were our Rolex watches – the Oyster Perpetual Rolex Deepsea model."[3] Although the watches were very different, echoes of the Royal Navy's early test dives, Dimitri Rebikoff's Mediterranean adventures, and Jacques Piccard's bathyscaphic escapades could easily be heard over the intervening decades.

Having devised the technical solution to keeping a mechanical timepiece functioning at extreme depth, it took the technical department of Rolex a little less than five weeks to develop and produce the ultimate Rolex Deepsea. The Deepsea Challenge was to be the watch that returned Rolex to a place it had last visited when Wilsdorf was alive: Challenger Deep in the Pacific Ocean, the deepest point of the planet's surface under the waves.

This forbidding section of the Mariana Trench south-west of Guam had last welcomed human visitors back in 1960 when Jacques Piccard and Don Walsh had made the perilous journey in the bathyscaphe *Trieste*, which had featured the famous hemispherical Deep Sea Special attached to the exterior of its hull.

Emmanuelle Périé-Bardout – Deepsea Under The Pole by Rolex expedition, 2010.

Deepsea Under The Pole by Rolex Expedition, 2010. Despite extreme conditions, the five Rolex Deepsea watches worked perfectly and allowed the team to track critical dive time.

Opposite: Rolex Deepsea, 2014, 904L steel, 44 mm, waterproof to 3,900 metres (12,800 feet). The hour markers and hands of the Chromalight display are filled with a luminescent material emitting a long-lasting blue glow – lasting up to twice as long as traditional materials.

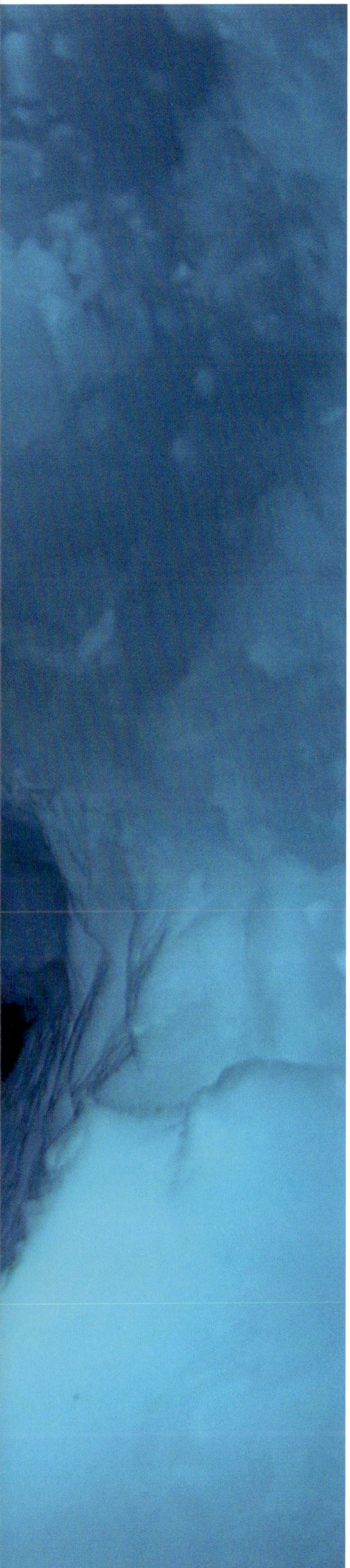

Since then, mankind had given the place a wide berth. The *Trieste* had become a historical curiosity, exhibited at the National Museum of the U.S. Navy. By comparison to Challenger Deep, the moon, which had borne the prints of 12 men's shoes, seemed like Times Square on New Year's Eve.

Half a century and more was to pass before film-maker and adventurer James Cameron began his solo descent into the terrifying abyss, wedged into the cramped cockpit of the *DEEPSEA CHALLENGER*. The director of *Titanic* and *Avatar* is a lifelong diving enthusiast for whom the Submariner enjoyed a talismanic significance long before he could afford one. "When you had one, it signified that you'd arrived – that you were a diver to be reckoned with. That for you, diving wasn't a hobby; it was a mission. And that your relationship with the ocean was deep and lifelong.

"It was a kind of wedding ring, a symbol of your marriage to the sea,"[4] he said, endowing the Submariner with the same quasi-mystical significance as the sacramental ring dropped into the Adriatic by the doges of Venice in the ceremonial *Sposalizio del Mare*, the millennium-old ritual of the Marriage of the Sea ceremony that signified La Serenissima's dominion over the ocean.

"When I finished *Aliens* in 1986, it was the first break I had in three years of work establishing myself as a film-maker. So, I went on a year-long dive safari, to apologise to the ocean for being away for so long. I had a little money, so I could afford the Rolex Submariner I'd always wanted. And, with my Submariner, I joined the ranks of the master divers I'd admired."[5]

It is therefore entirely fitting that for the dive of his life, attached to an external manipulator arm on that day, 26 March 2012, was an experimental Rolex watch: a Deepsea Challenge that had been designed and manufactured expressly for this single daring dive of nearly three hours down into the blackness at the bottom of the world.

In mid-2011, Rolex embarked on a study into the feasibility of producing a watch that could be made waterproof to a depth of 12,000 metres (39,370 feet). Highly involved calculations and modelling demonstrated that

Rolex Deepsea, 2014, Ref. 116660, 44 mm, 904L steel, waterproof to 3,900 metres (12,800 feet). Introduction of the D-blue dial, representing the colours of the deep.

GAS ESCAPE VALVE
10
20
30
40
50
28
SEA-DWELLER
SYSTEM

BECAUSE OF THE UNIQUE CONDITIONS, ALMOST EVERY ASPECT OF THE WATCH REQUIRED A NEW DESIGN.

the principles of the Ringlock system could be adapted to survive the most intense pressures and demanding conditions to be found beneath the world's oceans.

Rolex says its preparedness "demonstrated the brand's capacity to anticipate future needs"; however, it could be argued that the culture of Rolex ensured that it was not merely able to anticipate future needs, but to imagine situations so extreme as to be almost hypothetical, then devise the solutions to overcome them. And in order to create exceptionally robust timepieces, Rolex needed to devise exceptional ways of testing them – even if it required bespoke equipment that far exceeded accepted industry parameters.

Typical was the decision taken in 2004 to equip a test tank capable of testing the water resistance of watches to a guaranteed depth of 12,000 metres (39,370 feet). This was the extreme of hypothetical abstractions because these conditions do not exist in nature. There is no ocean floor that deep. Unsurprisingly this tank, quickly known by the soubriquet "Marianne" (alluding to the eponymous trench), was unique. Its true value would only become apparent in 2012, when Rolex, partnering with the National Geographic Society, joined the Deepsea Challenge and, incredibly, just 53 days before James Cameron's dive to the deepest point of the Pacific Ocean, work began on the watch.

The plan called for the preparation of not one but three watches for the dive, as well as an additional one that would be used solely for communication.

With initial calculations validated, detailed testing began. The issue that dominated those early days of this absurdly accelerated product launch was the dimensioning of the watch in order to arrive at the optimal overall compromise, as the official account of the project explains: "increasing the size of the case components would bring greater robustness but greater stress as well. The water pressure exerted on the watch would increase in proportion to its surface area."[6] Technical decisions were made in a matter of hours. Crucial to the success of the undertaking was the crystal, which would need to be 14.3 mm thick to be able to withstand the 1,500-bar (17-tonne pressure) test; but the shaping and geometry of the crystal was just as important as the thickness of material and "the exact specifications of the crystal were needed very quickly in order to deliver it within the one-week deadline. Briefed in the morning, the design engineers were able to come up with the dimensions the same evening, after an intense day of work on calculations and modelling."[7]

Just as the crystal would be of monumental proportions, so too the compression ring, which would enfold the movement in its protective embrace, was remarkably thick and solid. "This width determined the diameters of the crystal, of the titanium case back and of the 51.4-millimetre middle case. The design engineers concentrated on the technical characteristics of the heart of the piece, comprising the support ring, the crystal and the case back, which together ensure water and pressure resistance, as well as the crown. The rest of the case and bracelet, including the middle case and rotatable bezel, was rescaled from the Rolex Deepsea."[8]

By this point, on 10 February 2012 (44 days before the dive), the technical parameters had been established, but Rolex wanted to equip the expedition with more than a working prototype, which is what the original watch of 1960 had been. In that year, it had been sufficient to create a timepiece that could be attached to a bracelet and withstand the ultimate deep dive; it would have been too much to expect that this piece of equipment could wear like a normal watch as well, although that said, the original Deep Sea Special watch did subsequently appear in a very groovy magazine fashion shoot during the Swinging Sixties.

This time, Rolex aimed to do more than merely create an impregnable case in which to convey a mechanical watch movement to the bottom of the world; it wanted a proper watch, more precisely a watch in the familiar Submariner idiom.

Sworn to secrecy, one of the firm's top designers was inducted into the project. His first reaction was shock at what he politely described as the unusual proportions; his second, dismay at the deadline he was given: "Just half a day to readjust the appearance of the case and another half a day to design the markings and engravings on the dial and the flange."[9]

In order to avoid the top-heavy appearance that the gargantuan crystal could impart, it was decided to increase the thickness of the middle case to sheathe the sides of the crystal and minimize its dominance of the design. Further balance was added by increasing the thickness of the case back by a couple of millimetres and the diameter of the crown was enlarged to 10 mm.

Within a week, the first crystals were ready to be tested in the 1,500-bar Marianne tank. Ten crystals were tested.

The engineers held their breath. If the test failed, the project stopped there and then.

Relief and jubilation combined when all but one of the crystals emerged intact. At each stage of the process, the engineers and designers would encounter difficulties thrown up by the exaggerated scale of the design. For example, even the largest rod of 904L Oystersteel was of an insufficient diameter to be machined in the usual manner and had to be cut "lengthwise to obtain a rolled plate. This provided a large enough surface area to craft the middle case."[10]

Enlarged components and different methods of manufacture required new tools, and those working on the Deepsea Challenge "would design more tools than components for this experimental model."[11] Unsurprisingly, tool manufacturing was also an in-house skill, and the necessary items were swiftly produced by the prototyping department.

Because of the unique conditions, almost every aspect of the watch required a new design: whether the dial or the Cerachrom bezel insert, everything had to be appropriately proportioned.

What impresses most about this project is not necessarily that Rolex was capable of creating a watch to withstand the immense pressures at the bottom of the Challenger Deep – it had already accomplished that in 1960 – but rather that the watch that would repeat the journey in the 21st century would be as close in every possible respect to standard production models manufactured in their thousands.

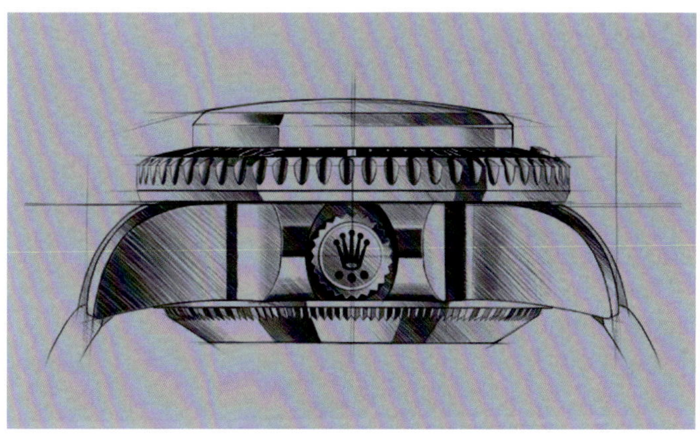

Sketch of the Deepsea Challenge experimental divers' watch.

Following pages, left: Making of the Deepsea Challenge. *Centre:* Exploded view of the watch case of the Deepsea Challenge and its main components. *Right:* Engraving of the flange of the Deepsea Challenge, 2012.

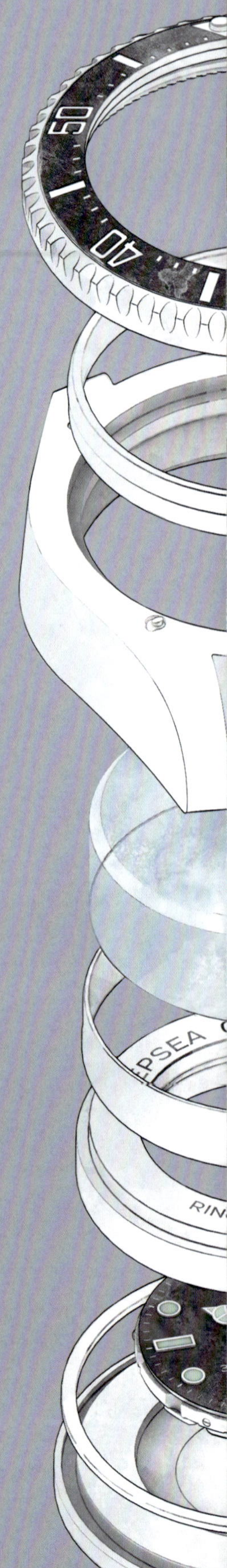
50
40

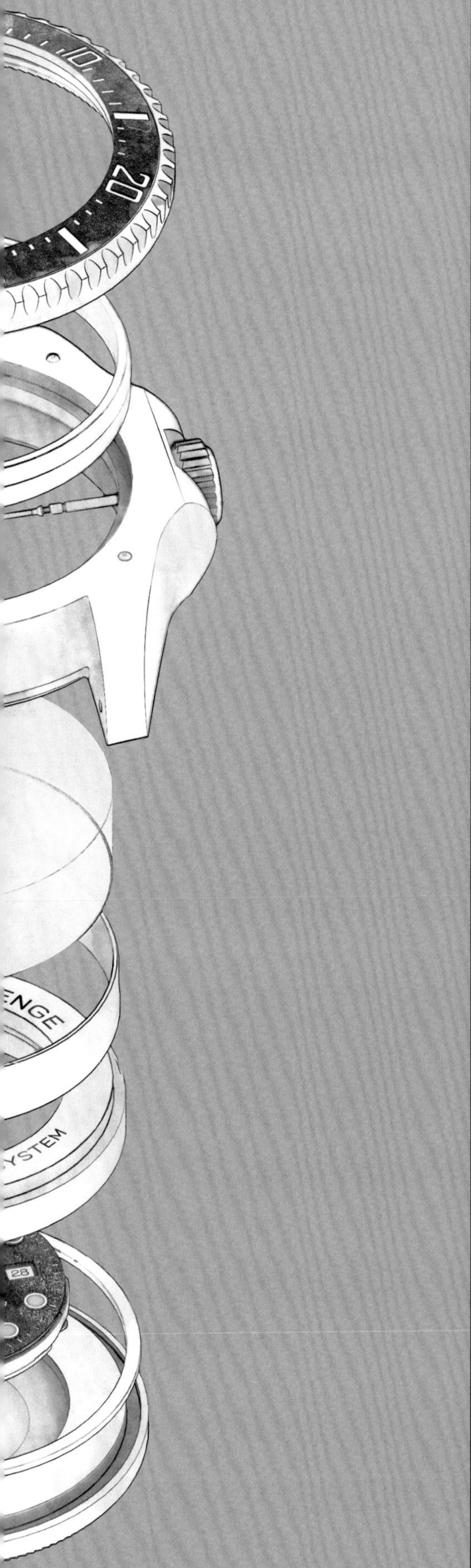

RING LOCK SYSTEM
ROLEX 2012 03

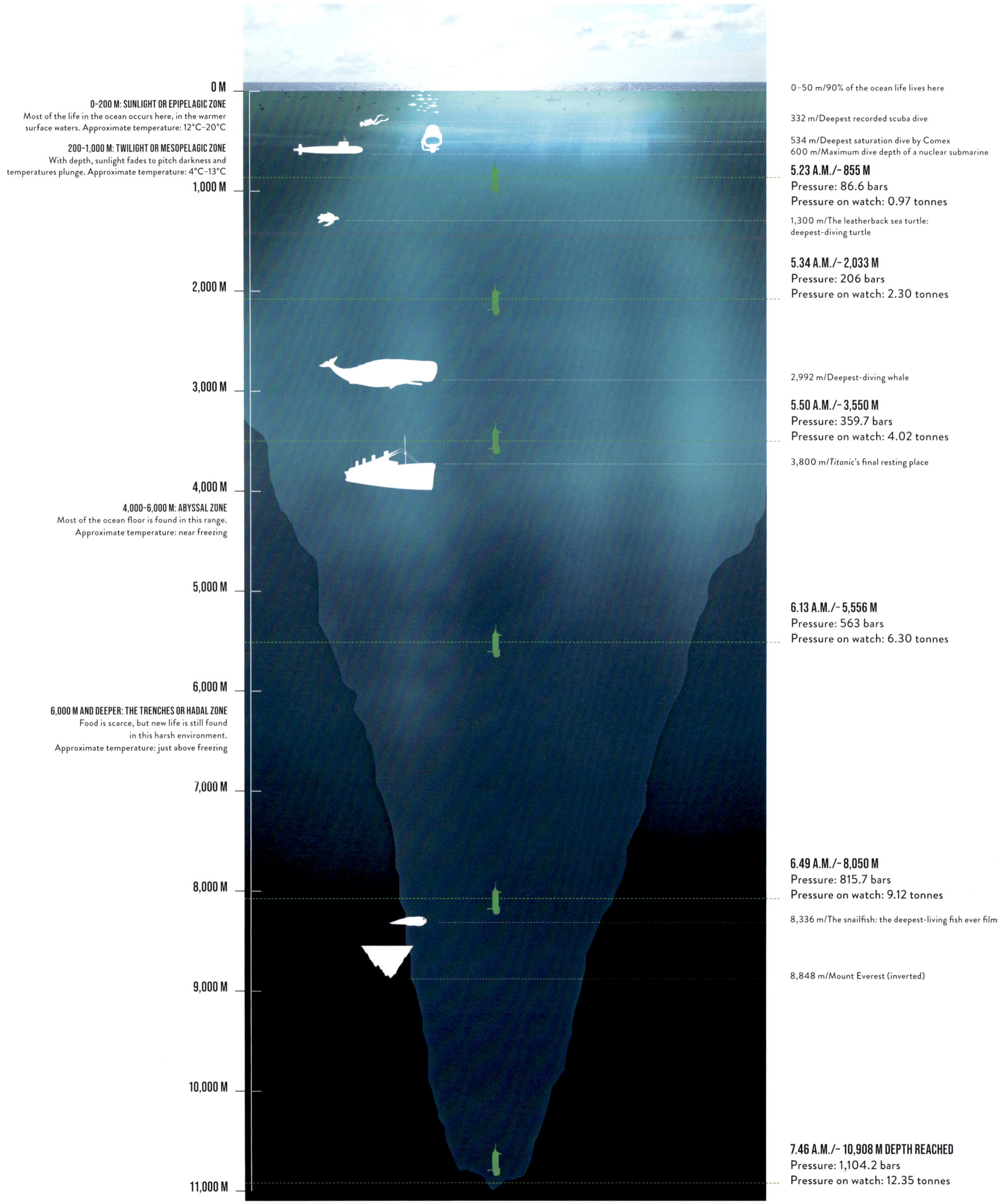
0 M
0–200 M: SUNLIGHT OR EPIPELAGIC ZONE
Most of the life in the ocean occurs here, in the warmer surface waters. Approximate temperature: 12°C–20°C
200–1,000 M: TWILIGHT OR MESOPELAGIC ZONE
With depth, sunlight fades to pitch darkness and temperatures plunge. Approximate temperature: 4°C–13°C
1,000 M
2,000 M
3,000 M
4,000 M
4,000–6,000 M: ABYSSAL ZONE
Most of the ocean floor is found in this range.
Approximate temperature: near freezing
5,000 M
6,000 M
6,000 M AND DEEPER: THE TRENCHES OR HADAL ZONE
Food is scarce, but new life is still found in this harsh environment.
Approximate temperature: just above freezing
7,000 M
8,000 M
9,000 M
10,000 M
11,000 M
0–50 m/90% of the ocean life lives here
332 m/Deepest recorded scuba dive
534 m/Deepest saturation dive by Comex
600 m/Maximum dive depth of a nuclear submarine
5.23 A.M./– 855 M
Pressure: 86.6 bars
Pressure on watch: 0.97 tonnes
1,300 m/The leatherback sea turtle: deepest-diving turtle
5.34 A.M./– 2,033 M
Pressure: 206 bars
Pressure on watch: 2.30 tonnes
2,992 m/Deepest-diving whale
5.50 A.M./– 3,550 M
Pressure: 359.7 bars
Pressure on watch: 4.02 tonnes
3,800 m/*Titanic*'s final resting place
6.13 A.M./– 5,556 M
Pressure: 563 bars
Pressure on watch: 6.30 tonnes
6.49 A.M./– 8,050 M
Pressure: 815.7 bars
Pressure on watch: 9.12 tonnes
8,336 m/The snailfish: the deepest-living fish ever film
8,848 m/Mount Everest (inverted)
7.46 A.M./– 10,908 M DEPTH REACHED
Pressure: 1,104.2 bars
Pressure on watch: 12.35 tonnes

The Deepsea Challenge was inspired by the experimental watch that was attached to the exterior of the *DEEPSEA CHALLENGER*, piloted by James Cameron to the bottom of the Mariana Trench on 26 March 2012. The submersible was subjected to extreme pressures during its descent, depicted in the illustration opposite.

"I FELT LIKE, LITERALLY IN THE SPACE OF ONE DAY, I HAD GONE TO ANOTHER PLANET AND COME BACK."

JAMES CAMERON

The headlong pace of modelling, testing and construction notwithstanding, adherence to the visual norms of standard Rolex production was maniacal. At 10 mm in diameter, the new watch's crown did not fit the dies for stamping the Rolex Crown emblem on the flat face; instead, once the correct proportions had been decided on, the familiar five-pointed crown and even the three dots that are the visual shorthand for the Triplock winding crown were laser-engraved.

After assembly and final checks, the three Deepsea Challenge watches were entrusted to the team who would hand-carry the watches to Guam and supervise their installation on the *DEEPSEA CHALLENGER* vessel.

Compared to the drama of the watches' conception, design and construction, the dive itself was straightforward. At 5.23 a.m. on 26 March 2012, the *DEEPSEA CHALLENGER* was lowered into the waters of the Pacific Ocean with three Rolex watches attached to its exterior. Two hours and 23 minutes later, Rolex returned to the bottom of the Challenger Deep. The *DEEPSEA CHALLENGER* remained on the seabed for about three hours before beginning its journey back to the surface.

Also on the support vessel was Mariana Trench veteran Don Walsh: "That was a grand moment, to welcome him to the club," Don Walsh told *National Geographic* that day. "There were only three of us in it, and one of them" – late Swiss engineer Jacques Piccard – "is dead. Now it's just Jim and myself."[12]

James Cameron had returned a changed man: "I felt like, literally, in the space of one day, I had gone to another planet and come back."[13]

However, for the watches, it was almost as if they had never left the Rolex headquarters in Geneva. These mute mechanical witnesses to this historic moment emerged from the Pacific Ocean as they had entered it: intact, unmarked and functioning perfectly.

"The Deepsea Challenge was the reliable companion throughout the dive, it was visible on the sub's manipulator arm and working precisely at more than 10,000 metres down at the bottom of Challenger Deep," said James Cameron on completion of his historic dive. "It's a tremendous example of engineering know-how, and an ideal match for the *DEEPSEA CHALLENGER* submersible."

James Cameron after his dive to the bottom of the Mariana Trench.

Rolex's commitment to undersea exploration led the brand to take an active part in film-maker and explorer James Cameron's expedition in 2012 in partnership with the National Geographic Society. The experimental Deepsea Challenge watch was part of the journey, affixed to the articulated arm of the *DEEPSEA CHALLENGER* submersible.

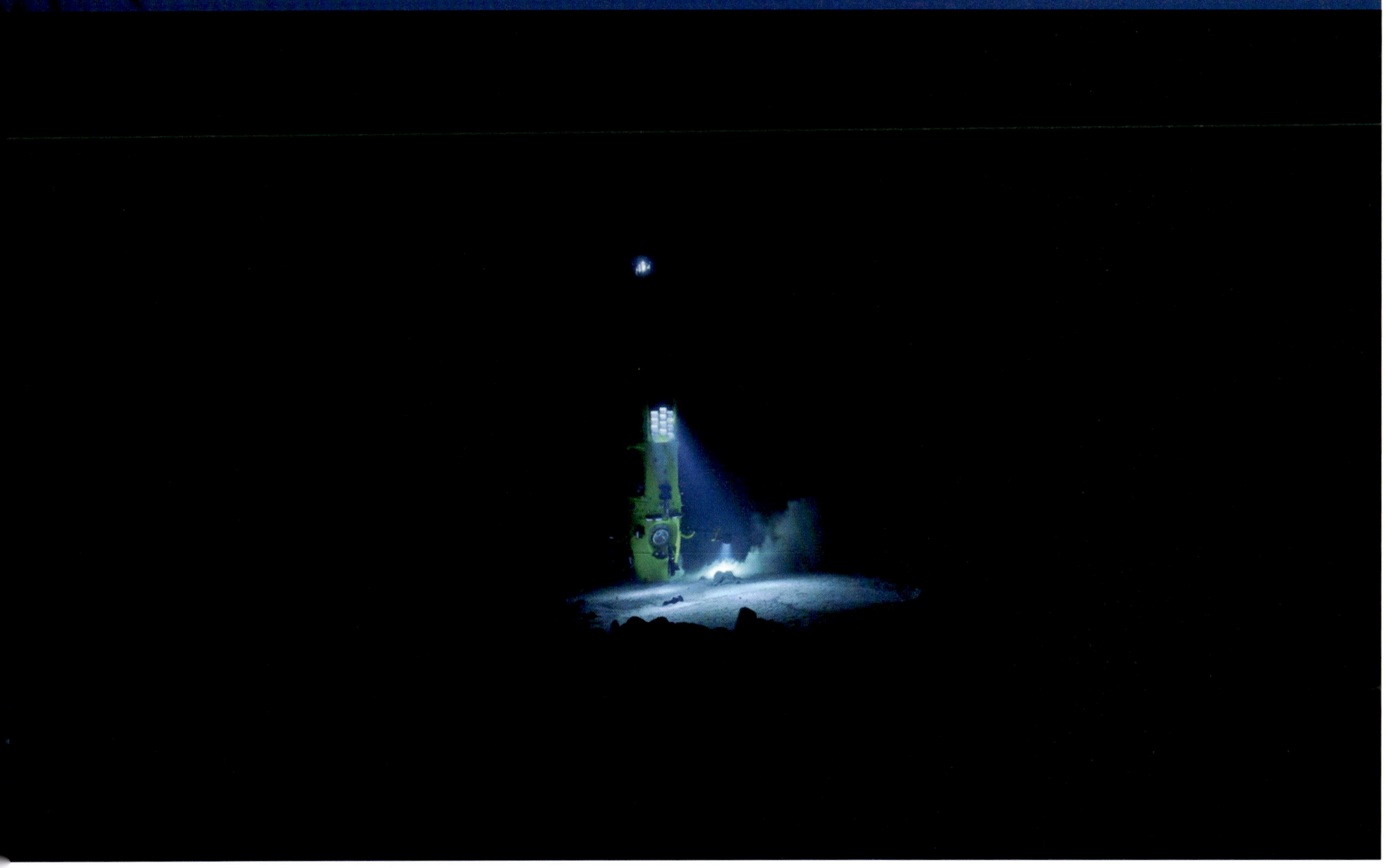

MARIANA TRENCH
ROLEX
26-03-2012
23-01-1960

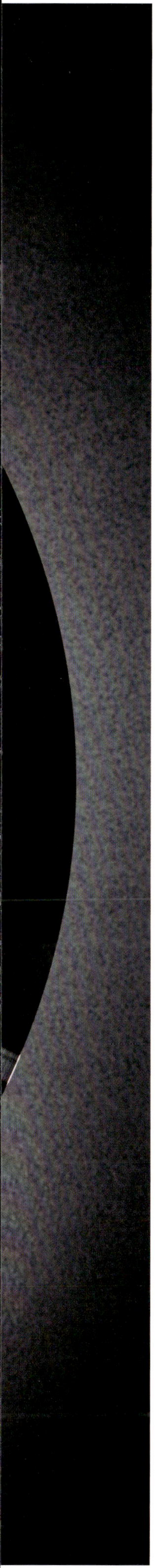

Ultra-high-pressure tank to test the waterproofness of the Deepsea Challenge, launched in 2022.

Opposite: A decorative plate atop the waterproofness testing tank for the Deepsea Challenge featuring similar engravings to the case back of the watch.

Top row: Rolex Testimonees Cristina Mittermeier, Titouan Bernicot and Paul Nicklen.

Bottom row: Rolex Awards for Enterprise Laureates Brad Norman, Luiz Rocha and Emma Camp.

PERPETUAL PLANET INITIATIVE: A FAMILY OF OCEAN HEROES

The Submariner has always been the watch of heroes: naval heroes, heroes who risked their lives exploring the underwater world and, of course, those of the cinema screen. Now, a quarter of the way through the 21st century, a new generation of heroes is wearing the Submariner. They are just as fearless in the protection and preservation of the ocean environment as the marine explorers who earlier embraced this storied timepiece.

For Mexican-born photographer, marine biologist and conservationist Cristina Mittermeier, her Submariner is much more than her horological companion to the remotest places on the planet: it is the physical manifestation of her mission in life. "With my Submariner, I am never wondering only about the time; I am reminded of the imperative to protect our ocean and the growing family of heroes who are now part of Rolex's Perpetual Planet Initiative."

Watching the seconds hand make its minute-long circuit on her watch's dial, Mittermeier reflects on how the sands of time are running through the hourglass of our planet's survival. For her, the Submariner is not just an indicator of the hours, minutes and seconds that pass: it represents a standard to live up to, as she toils tirelessly to preserve the marine world that was once innocently enjoyed before growing understanding brought awareness of the wounds being inflicted on our world by unfettered industrial development and the gathering pace of the climate emergency.

With knowledge came a sense of responsibility and, for Mittermeier, the duty to act. Although she studied biochemical engineering in marine sciences, she soon understood that she could not achieve what she wanted within the abstract academic confines of the laboratory, where analysis subsumes emotion and research replaces passion. She took off her lab coat and set out to kindle in others the passion that burned within her. "I left biology in the hope of becoming a voice for the ocean and its peoples. To tell its story with a camera."[1]

"YOU MIGHT THINK YOU'RE JOURNALISTS BUT YOU'RE ALSO CONSERVATION PHOTOGRAPHERS IN THE SERVICE OF TRYING TO MAKE A BETTER PLANET."

CRISTINA MITTERMEIER

Shortly after the turn of the millennium, in 2005, Mittermeier signified important change in the nature of her vocation when she founded the International League of Conservation Photographers. The term "nature photographer" or "wildlife photographer" was no longer adequate to describe the type of work that she and her peers were undertaking. Photography was no longer solely a means of documenting the natural world. It was instead a form of activism and advocacy, a powerful medium that could make others aware of the planet in peril, with images that were at once tragic and beautiful, simultaneously inspiring and harrowing.

In skilled hands, argues Mittermeier, a camera is not a passive instrument registering images but a versatile and effective weapon in the fight to save our planet, and to preserve the world and the wildlife she loves to photograph. "My goal is to give voice to the natural world and I point my camera wherever it's needed – where imagery and storytelling have the potential to create positive change."

Mittermeier urged her peers to take their cameras to the remotest corners of the world, be it the Amazon or the Antarctic, and then carry the message to challenge the complacency of Western civilization. "We live in an economic paradigm based on growth that is completely incompatible with the planet's finite resources," she says. "To me the goal is to have a voice to protect the natural world, that's what I care about."[2]

Among those who heeded her call to arms was Paul Nicklen, a Canadian photographer and marine biologist who had spent his youth living among the Inuit community of Kimmirut inside the Arctic Circle on Baffin Island. The child who played under the aurora borealis and hiked to the top of mountains grew into a highly accomplished photographer with an impressive record of stories published in *National Geographic*.

Nicklen found himself swept up by Mittermeier's persuasive energy. "Cristina said, 'You might think you're journalists but you're also conservation photographers in the service of trying to make a better planet'."[3] These words were to change his life. He could not unhear what she had said, and from then on, he thought of himself not as a photographer of wildlife and nature, but "as a war photographer, going into battle for nature".[4]

Opposite and following pages: Together with photographer Paul Nicklen, Cristina Mittermeier founded SeaLegacy, an organization that uses world-class visual storytelling to inspire action to restore the health of the oceans.

Nauticam

Paul Nicklen spent over 20 years as an assignment photographer for the National Geographic Society. He returns from his dives with breathtakingly beautiful photographs, images that alert us to the need to protect the ecosystems of these regions.

He became her partner in life as well as ideology: "My work illustrates how human actions and climate change are derailing ecosystems."

In 2014, frustrated that preserving the ocean was the most poorly funded area of conservation, Mittermeier and Nicklen founded the non-profit organization SeaLegacy, pledging to use the latest "strategic communications at the intersection of art, science and conservation to protect and rewild the ocean for the benefit of biodiversity, humanity and climate within our lifetimes".[5]

The message was to celebrate and preserve the oceans. The medium, the lingua franca of photography, was the same as in the distant days of Dimitri Rebikoff, whose pioneering work in subaqueous photography and lighting had first brought the colourful, glorious and apparently infinite abundance and diversity of the marine environment to the attention of humankind. But over three quarters of a century, the means of disseminating these images and broadcasting their message of imperilled wonder has moved beyond the slideshows, film projectors and colour magazines of the mid-20th century. With the passage of time, the crisis hastened and a more urgent method of carrying the message was needed to supplement traditional forms. "With social media, rather than spend a year shooting for a magazine, I could put stuff out there and I quickly got millions of followers and you realize you're your own news network and you're shooting daily for an audience," explains Nicklen.[6]

And, in addition to the immediacy of social media, the increasing acceptance of photography as a culturally legitimate branch of the fine arts elevates Nicklen's subjects to the timelessness of great art.

Nicklen's work captures the eternity of a candid moment, a split second. *Majesty Surfacing*, a work from 2012, depicts the rivulets of water streaming from the coat of a grizzly bear as it breaks the surface of a river in the northern Yukon. This near magical ability creates a photographic essay in form. *Ephemeral Palace* (2012) depicts a fantastical sprawling structure of ice: all soaring towers, powerful buttresses and vaulting cathedral-like arches, blue against slate skies and anthracite waters. Sculpted by time and climate from million-year-old ice, its majestic pomp is freighted with poignancy as, drifting slowly away from Antarctica, it begins its doomed journey until it eventually becomes one with the waters of the world's oceans.

Paul Nicklen uses his photography of the polar regions and the oceans to document the magnificence of the wild world as well as its fragility in the face of climate change.

"WHEN I AM WORKING UNDERWATER, TRYING TO FIND THE KIND OF TOUGH CORALS WE NEED TO REPOPULATE THE REEFS, MY SUBMARINER IS ALWAYS ON MY WRIST."

TITOUAN BERNICOT

Even though Nicklen was born in 1968 and has spent his working life as a photographer, carrying off his first World Press Photo Award in 2004, the freshness of his work remains undimmed as he continues to marvel at "the power of photography to grip a viewer with nothing more than a single moment".[7]

"Young people are facing eco-anxiety," says Nicklen. "Doing nothing only increases your anxiety and the only antidote is to take action. If we rely on governments and hope we're going to elect the right people to save our planet, we're doomed. Once you have knowledge about what's happening to the planet, doing nothing isn't an option."

Nicklen, Mittermeier and their peers have inherited the struggle from an earlier generation of the Rolex family. In 2021, Mittermeier was awarded the Seattle Aquarium Sylvia Earle Medal that carries the name of a famed Rolex Testimonee. With her dissertation on marine plant life, *Phaeophyta of the Eastern Gulf of Mexico*, Earle was awarded her PhD in 1966, Mittermeier's birth year.

And it is through Mittermeier, Nicklen and their like that the sense of responsibility is being passed to a generation born at the turn of the century, for whom the symbiotic and inalienable bond between humankind and the ocean that covers over 70 per cent of the world's surface is sacred.

Only by understanding the world's ocean as a single entity – to use Mittermeier's words: "a unified, planetary, wide, global body of water"[8] – do we begin to appreciate the interconnected nature of life on our fragile planet.

In the interview she gave to mark her acceptance of the prestigious Sylvia Earle Medal, Mittermeier made mention of Titouan Bernicot, a young man born on a remote Polynesian pearl farm. "At age 16, he realized that the coral reefs in French Polynesia were dying and decided to do something to help. He founded Coral Gardeners, a non-profit that educates the public about the importance of coral reefs. Now 23 years old, Titouan works alongside his fellow Coral Gardeners to restore the reef on their home island of Mo'orea. Spending time with them and my team at SeaLegacy reinvigorated my own sense of purpose and filled me with hope."[9]

Rolex, too, was inspired by the example of this young man and, since 2021, the Rolex Perpetual Planet Initiative has supported Coral Gardeners in its work, training young

Coral Gardeners, an NGO founded by Titouan Bernicot and led by young people, restores reefs using resilient corals in French Polynesia.

Polynesians in the skills and techniques necessary for propagating corals, tens of thousands of which have been transplanted since the organization's foundation in 2017.

Bernicot is now busy building up his organization with his Submariner on his wrist. "I spend most of my time raising awareness for ocean conservation and trying to save coral reefs from the effects of climate change through Coral Gardeners. When I am working underwater, trying to find the kind of tough corals we need to repopulate the reefs, my Submariner is always on my wrist. I also wear it around the world when I am asked to speak about what we do. This Rolex timepiece will accompany me during the many expeditions planned for our next chapter as we scale up reef restoration around the globe. It will remind me to keep pushing boundaries for the future of our blue planet."

Through the actions of its extraordinarily dedicated Testimonees and Rolex Awards for Enterprise Laureates such as marine biologists Brad Norman, Emma Camp and Luiz Rocha, as well as the projects it supports through the Perpetual Planet Initiative, Rolex ensures the transmission, from generation to generation and from century to century, of the crucial message that action needs to be taken now to preserve our water-covered planet for the future.

Since work began on the timepiece that would become the world's most famous diving watch, humanity's relationship with the seas has changed utterly. Exploration and exploitation have given way to conservation and restoration. However, one thing remains the same: the benign presence of the Submariner, a familiar expression of the perpetual values of precision and dependability, a steadfast and practical mechanical friend in a changing and uncertain world.

Nicklen puts it well: "When I am diving, there is no room for error. Filming underwater, surrounded by high-tech systems like rebreathers and state-of-the-art cameras, I gravitate towards simple devices whenever I can. That is why I like the Rolex Submariner watch – it gives me the time and backup bottom-time tracking when diving. Simplicity, quality and reliability are the most important attributes of this watch."[10]

The Coral Nurture Program was co-founded by Rolex Awards for Enterprise Laureate Emma Camp. More than 85,000 corals have been transplanted by the British marine biologist and her team to 124 nurseries over 30 sites on the Great Barrier Reef, helping to rebuild its resilience.

mares

HOLLIS
TEMPERED CRYSTAL CLEAR LENS
TEMPERED CRYSTAL CLEAR LENS
HYDRA
XS SCUBA

Submariner, 2020,
Ref. 124060, 41 mm, Oystersteel,
waterproof to 300 metres (1,000 feet).

Opposite: Luiz Rocha, a Rolex Awards for Enterprise Laureate in 2021, photographed in the Maldives wearing the Submariner Date.

EPILOGUE

"By returning humans to the so-called hadal zone – the ocean's deepest level, below 20,000 feet (6,000 metres) – the Challenger Deep expedition may represent a renaissance in deep-sea exploration."[1]

This was *National Geographic's* observation of mankind's return to the deepest point beneath the sea after an absence of 52 years. "The critical thing is to be able to take the human mind down into that environment," explained marine geologist and expedition member Patricia Fryer, "to be able to turn your head and look around to see what the relationships are between organisms in a community and to see how they're behaving – to turn off all the lights and just sit there and watch and not frighten the animals, so that they behave normally."[2]

Much had changed in the half century since Don Walsh, Jacques Piccard and Rolex had ventured into the dark unknown of the Challenger Deep. In 1960, the undertaking had been a military mission conducted by the U.S. Navy. With the Space Race well underway and the Bay of Pigs as well as the Cuban Missile Crisis in the near future, the descent had become an exercise in militaristic patriotism. The *Trieste* touched down on the sediment at the bottom of the Challenger Deep almost as a conqueror. Flags were taken to the seabed by Walsh and Piccard, recalling the European explorers of preceding centuries, who had travelled to inhospitable, uncharted parts of the world to claim them on behalf of their emperors, monarchs and governments. After its astonishing voyage, the *Trieste*, property of the U.S. Navy, had been hauled through Washington D.C., paraded along the streets on a pantechnicon emblazoned with the slogan "Navy Conquers Inner Space". The achievement may have been made possible only by the latest scientific advances of the 20th century, but the manner of its celebration was rooted in the exultant militarism of the triumphs Ancient Rome had granted those of its generals who had been successful in battle.

Now, in the 21st century, humankind had returned as a respectful visitor with a multidisciplinary scientific mission and the goal of learning and inspiring rather than conquering. The presence of Rolex was more than a nostalgic nod to the earlier dive, but rather a demonstration that Rolex continued to be humanity's faithful and dependable companion in the ongoing exploration of the planet; exploration that was taking place in a new spirit. The remote corners of the world seldom visited by humankind were no longer viewed as difficult challenges needing to be overcome and subdued, but rather environments that needed to be understood, respected and preserved.

James Cameron, who had studied science before becoming a film-maker, declared that he had made the descent in order to unravel the riddles of life on the planet we share with so many other species and that we still understand only imperfectly. But he also alluded to a larger mission: he hoped that his voyage to the bottom of the world would awaken the curiosity of a younger generation and inspire them to explore, understand, protect and perpetuate the planet. His journey to the very bottom of the sea certainly inspired Rolex to deepen, if you will excuse the pun, its commitment to the world under the waves. One of the most surprising manifestations of that constant, one could say perpetual, commitment came at the end of 2022 when, in a move that took the watch industry by surprise, Rolex launched the Oyster Perpetual Deepsea Challenge. Much more than another deep-dive watch, it was described simply – and boldly – as the ultimate divers' watch. Rolex had developed a timepiece with the immense pressure-resistance of the experimental piece that had accompanied Cameron to the bottom of the ocean, but that was also genuinely wearable in the world above water: it had shrunk by 1.4 mm in diameter to 50 mm and 5.2 mm in height to 23.3 mm; moreover, it was about 30 per cent lighter.

As with all Rolex divers' watches, it is tested to an additional 25 per cent beyond the depth stated on the dial, meaning that it is capable of resisting pressure at the totally abstract depth of almost 14 kilometres underwater – far, far deeper than the deepest place on our planet.

Rolex was the first to make watches to descend to the deepest depths, and this new watch extended that supremacy.

Submariner Date, 2020,
Ref. 126618 LB, 41 mm,
18 ct yellow gold, waterproof to
300 metres (1,000 feet).

Coral reefs support 25 per cent of marine biodiversity. Rolex's Perpetual Planet Initiative is working with partners like Rolex Awards for Enterprise Laureate Emma Camp, who are finding ways to help them build resilience in the face of climate change.

ORIGINAL GAS ESCAPE
ROLEX
SEA-DWELLER

Having already made a watch that could function at maximum depth in 1960 and then again in 2012 for Cameron, the next challenge was to produce such a watch in series.

This not only required specialized design and innovative use of materials, but new validation procedures. The equipment conducting the barrage of tests that need to be completed before a watch can carry the coveted designation Superlative Chronometer was incapable of accommodating the Deepsea Challenge. Thus, a bespoke testing facility was set up to run in parallel with the automated Superlative Control testing line. But whereas the standard testing procedures are carried out behind toughened plate glass, albeit 30 metres (100 feet) below Rolex's Geneva headquarters, the apparatus for testing the waterproofness of the Deepsea Challenge is so specialized that it is kept out of sight in an area requiring key card entry.

Unsurprisingly, the most demanding criterion to fulfil is the pressure testing. Superlative Control testing offers two levels of pressure: middle pressure (MP), approximately 280 bars, and high pressure (HP) of 630 bars for the regular Rolex Deepsea. For the Deepsea Challenge, a third designation was required: ultra-high pressure (UHP), able to generate a massive 1,750 bars.

One of the reasons that these watches are made in comparatively small quantities is that the capacity of the tank in which they are tested is just three and a half litres. "Tank" may potentially be a misleading term inasmuch as the device is in fact a tube about 10 centimetres (4 inches) wide and more than 40 centimetres (16 inches) deep. In order to carry out UHP testing, the watches are hand-loaded into a slotted frame that stacks a maximum of eight watches, and then lowered into the water. With the watches fully immersed, the tank is sealed by a system reminiscent of a sophisticated bank vault.

So far so unusual, until you realize that the machinery built to test a handful of wristwatches weighs 1,500 kilograms, the same as a medium-sized family car. It needed to be specially built as such machinery did not exist. The pressure testing tank was designed, constructed and installed by Rolex's long-standing collaborator on deep diving research, Comex.

Deepsea Challenge, 2022, Ref. 126067, 50 mm, RLX titanium, waterproof to 11,000 metres (36,090 feet).

To make a prototype watch capable of withstanding such immense pressure as Rolex had done for Piccard and Walsh in 1960 and for Cameron in 2012 is impressive; to transfer these remarkable properties into a considerably more compact, lighter timepiece to be made in series and offered to the public is astonishing.

The breakthrough of substantial weight loss had been achieved through the use of a grade 5 titanium alloy called RLX titanium. The Deepsea Challenge was made of titanium so that in spite of its size, it would remain comfortable to wear. The logic of this decision is unassailable, but it represented a significant cultural shift: the Deepsea Challenge was the first Rolex with a case and bracelet made from titanium. But it was not to be the last. Just a few months later at the Watches and Wonders fair in Geneva in March 2023, Rolex announced a 42 mm Yacht-Master in titanium and, as with the Deepsea Challenge, the motivation behind this choice was technical rather than aesthetic.

Just as it mastered the depths, so, too, has Rolex mastered titanium, and its use in the Yacht-Master 42 range was fitting given Rolex's involvement with SailGP, the exciting new sailing championship in which crews compete in specialized, high-performance foiling catamarans that almost seem to fly just above the waves. In this exhilarating sport, weight is the enemy, every excess gram is trimmed, and as a technological material with similar properties to steel but without its weight, RLX titanium is the appropriate material to use in a Professional watch for SailGP.

CONCLUSION

It used to be said that "Britannia Rules the Waves", but that was before the Rolex Submariner existed. As Jean-Frédéric Dufour, Rolex's CEO, puts it: "The watch mastering waterproofness is Rolex ... it's part of history." It is also very much part of the present and the future. Whether skimming atop the sea in a highly specialized racing catamaran or diving to the deepest point on the planet, Rolex makes the appropriate wristwatch not just for those two extremes, but also for the waters in between. When considered in that light, the Submariner and its family of watches is truly one of mankind's most remarkable creations.

But as well as a feat of horological engineering, the Submariner also says something about Rolex's approach to watchmaking and, beyond that, the world at large.

"EVERY SINGLE BRAND HAS A DIVERS' WATCH. BUT NOBODY HAS A CONNECTION WITH THE SEA LIKE ROLEX HAS."

JEAN-FRÉDÉRIC DUFOUR, *Rolex CEO*

For Rolex watchmaking, and life for that matter, is not about the facile, hollow triumph of breaking records and then moving on to another highly visible challenge. The Rolex approach is about long-term commitment and staying power. Long after the glare of public attention and the spotlight of marketing hype have moved on, Rolex remains unwavering in its commitment to improving and perfecting all its watches, whether the humble Oyster Perpetual or, in the case of the Submariner, a highly specialized Professional model. It is this commitment – not over years and decades, but from generation to generation – that makes Rolex ... well ... Rolex.

The Submariner's place in the world in the early decades of the 21st century is assured. Born out of the new needs of underwater exploration in the mid-20th century and then developed rapidly as the novel technologies of the aqualung and subaquatic filming opened up the world beneath the waves, the Submariner was first tested and developed by a generation of pioneers, and although these remarkable characters are no more, they bequeath those who follow them with the spirit of wonder and curiosity out of which the Submariner emerged.

Seventy years later, that same spirit remains as much a part of the world of the Submariner as ever. Today, the decades of exploration and discovery to which this watch has contributed, and the momentous pages in history that it has written, have helped us understand both the value and jeopardy of our Earth, and in turn have given new meaning to the familiar face of the world's most famous divers' watch. Each new Submariner produced and worn today is a tacit commitment to the struggle for what Rolex calls the Perpetual Planet.

Jean-Frédéric Dufour puts it very simply: "Every single brand has a divers' watch. But nobody has a connection with the sea like we have."

Rolex Deepsea, 2024, Ref. 136668 LB, 44 mm, 18 ct yellow gold, waterproof to 3,900 metres (12,800 feet).

THE SUBMARINER – EVOLUTION OF AN ICON

SUBMARINER

Submariner watches epitomize the historic link between Rolex and the world of underwater exploration. Created in 1953, the Oyster Perpetual Submariner was the first divers' wristwatch waterproof to a depth of 100 metres (330 feet). In 1969, Rolex unveiled a second Submariner model, with a date function: the Oyster Perpetual Submariner Date. The waterproofness of both models was later increased to 300 metres (1,000 feet). Thanks to their renowned robustness, the Submariner and Submariner Date also moved beyond the marine depths, becoming the watches of choice for a wide range of sports and other activities.

SEA-DWELLER

Launched in 1967, the Oyster Perpetual Sea-Dweller was created to accompany the pioneers of deep-sea diving. This watch is the result of years of collaboration and experimentation with professionals and organizations specializing in undersea exploration. It is notably equipped with the helium escape valve, an innovation that has played a key role in underwater research since it first accompanied a new professional diving technique: saturation diving. Initially capable of descending to 610 metres (2,000 feet), the Sea-Dweller has, since 1978, been guaranteed waterproof to a depth of 1,220 metres (4,000 feet).

DEEPSEA

Watches in the Deepsea range demonstrate the unique expertise of Rolex in waterproofness and resistance to underwater pressure. Guaranteed waterproof to a depth of 3,900 metres (12,800 feet), the Oyster Perpetual Rolex Deepsea inaugurated, at its launch in 2008, the Ringlock system – a unique case architecture that allows the watch to withstand the pressure exerted at this extreme depth. Released in 2022, the Oyster Perpetual Deepsea Challenge also benefits from this exclusive construction. This watch, crafted from RLX titanium, is guaranteed waterproof to the phenomenal depth of 11,000 metres (36,090 feet). Equipped with all the innovations developed by the brand for its divers' watches, it embodies a perfect balance of superlative technical performance and wearer comfort.

SUBMARINER REF. 6204
Launch year: **1953**
Production quantity: **2,881**
Material: **STAINLESS STEEL**
Diameter: **38 MM**
Calibre: **9¾ A260**
Function: **HOURS, MINUTES, SECONDS**
Waterproofness: **100 M/330 FT**

SUBMARINER REF. 6200
Launch year: **1954**
Production quantity: **303**
Material: **STAINLESS STEEL**
Diameter: **38 MM**
Calibre: **10½ A296**
Function: **HOURS, MINUTES, SECONDS**
Waterproofness: **200 M/660 FT**

SUBMARINER REF. 6205
Launch year: **1954**
Production quantity: **810**
Material: **STAINLESS STEEL**
Diameter: **38 MM**
Calibre: **9¾ A260**
Function: **HOURS, MINUTES, SECONDS**
Waterproofness: **100 M/330 FT**

SUBMARINER REF. 6538
Launch year: **1954**
Production quantity: **2,800**
Material: **STAINLESS STEEL**
Diameter: **38 MM**
Calibre: **1030**
Function: **HOURS, MINUTES, SECONDS**
Waterproofness: **200 M/660 FT**

SUBMARINER REF. 6536
Launch year: **1955**
Production quantity: **5,350**
Material: **STAINLESS STEEL**
Diameter: **38 MM**
Calibre: **1030**
Function: **HOURS, MINUTES, SECONDS**
Waterproofness: **100 M/330 FT**

SUBMARINER REF. 5508
Launch year: **1958**
Production quantity: **9,000**
Material: **STAINLESS STEEL**
Diameter: **38 MM**
Calibre: **1520-1530**
Function: **HOURS, MINUTES, SECONDS**
Waterproofness: **100 M/330 FT**

SUBMARINER REF. 5510
Launch year: **1958**
Production quantity: **500**
Material: **STAINLESS STEEL**
Diameter: **38 MM**
Calibre: **1520-1530**
Function: **HOURS, MINUTES, SECONDS**
Waterproofness: **200 M/660 FT**

SUBMARINER REF. 5512
Launch year: **1959**
Production quantity: **17,338**
Material: **STAINLESS STEEL**
Diameter: **40 MM**
Calibre: **1530-1560-1570**
Function: **HOURS, MINUTES, SECONDS**
Waterproofness: **200 M/660 FT**

SUBMARINER REF. 5513
Launch year: **1962**
Production quantity: **151,449**
Material: **STAINLESS STEEL**
Diameter: **40 MM**
Calibre: **1520-1530**
Function: **HOURS, MINUTES, SECONDS**
Waterproofness: **200 M/660 FT**

SEA-DWELLER REF. 1665
Launch year: **1967**
Production quantity: **22,038**
Material: **STAINLESS STEEL**
Diameter: **40 MM**
Calibre: **1575**
Function: **HOURS, MINUTES, SECONDS AND DATE**
Waterproofness: **610 M/2,000 FT**

SUBMARINER REF. 5514
Launch year: **1969**
Production quantity: **1,618**
Material: **STAINLESS STEEL**
Diameter: **40 MM**
Calibre: **1520-1530**
Function: **HOURS, MINUTES, SECONDS**
Waterproofness: **200 M/660 FT**

SUBMARINER REF. 1680
Launch year: **1969**
Production quantity: **111,226**
Material: **STAINLESS STEEL**
Diameter: **40 MM**
Calibre: **1575**
Function: **HOURS, MINUTES, SECONDS AND DATE**
Waterproofness: **200 M/660 FT**

SUBMARINER REF. 1680
Launch year: **1969**
Production quantity: **5,765**
Material: **18 CT YELLOW GOLD**
Diameter: **40 MM**
Calibre: **1575**
Function: **HOURS, MINUTES, SECONDS AND DATE**
Waterproofness: **200 M/660 FT**

SUBMARINER REF. 5517
Launch year: **1977**
Production quantity: **539**
Material: **STAINLESS STEEL**
Diameter: **40 MM**
Calibre: **1520-1530**
Function: **HOURS, MINUTES, SECONDS**
Waterproofness: **200 M/660 FT**

SEA-DWELLER REF. 16660
Launch year: **1978**
Production quantity: **16,854**
Material: **STAINLESS STEEL**
Diameter: **40 MM**
Calibre: **3035**
Function: **HOURS, MINUTES, SECONDS AND DATE**
Waterproofness: **1,220 M/4,000 FT**

SUBMARINER REF. 16800
Launch year: 1979
Production quantity: 81,694
Material: STAINLESS STEEL
Diameter: 40 MM
Calibre: 3035
Function: HOURS, MINUTES, SECONDS AND DATE
Waterproofness: 300 M/1,000 FT

SUBMARINER REF. 16808
Launch year: 1979
Production quantity: 5,988
Material: 18 CT YELLOW GOLD
Diameter: 40 MM
Calibre: 3035
Function: HOURS, MINUTES, SECONDS AND DATE
Waterproofness: 300 M/1,000 FT

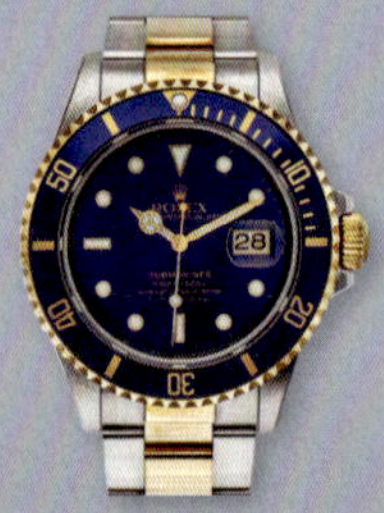

SUBMARINER REF. 16803
Launch year: 1983
Production quantity: 30,239
Material: YELLOW ROLESOR
Diameter: 40 MM
Calibre: 3035
Function: HOURS, MINUTES, SECONDS AND DATE
Waterproofness: 300 M/1,000 FT

SUBMARINER REF. 16610
Launch year: 1986
Production quantity: 646,613
Material: STAINLESS STEEL
Diameter: 40 MM
Calibre: 3135
Function: HOURS, MINUTES, SECONDS AND DATE
Waterproofness: 300 M/1,000 FT

SUBMARINER REF. 16613 LN
Launch year: 1986
Production quantity: 106,815
Material: YELLOW ROLESOR
Diameter: 40 MM
Calibre: 3135
Function: HOURS, MINUTES, SECONDS AND DATE
Waterproofness: 300 M/1,000 FT

SUBMARINER REF. 16613
Launch year: 1986
Production quantity: 296,784
Material: YELLOW ROLESOR
Diameter: 40 MM
Calibre: 3135
Function: HOURS, MINUTES, SECONDS AND DATE
Waterproofness: 300 M/1,000 FT

SUBMARINER REF. 16618
Launch year: 1986
Production quantity: 20,568
Material: 18 CT YELLOW GOLD
Diameter: 40 MM
Calibre: 3135
Function: HOURS, MINUTES, SECONDS AND DATE
Waterproofness: 300 M/1,000 FT

SEA-DWELLER REF. 16600
Launch year: 1986
Production quantity: 176,602
Material: STAINLESS STEEL
Diameter: 40 MM
Calibre: 3135
Function: HOURS, MINUTES, SECONDS AND DATE
Waterproofness: 1,220 M/4,000 FT

SUBMARINER REF. 14060
Launch year: 1989
Production quantity: 93,518
Material: STAINLESS STEEL
Diameter: 40 MM
Calibre: 3000
Function: HOURS, MINUTES, SECONDS
Waterproofness: 300 M/1,000 FT

SUBMARINER REF. 14060 M
Launch year: 2001
Production quantity: 130,622
Material: STAINLESS STEEL
Diameter: 40 MM
Calibre: 3130
Function: HOURS, MINUTES, SECONDS
Waterproofness: 300 M/1,000 FT

SUBMARINER REF. 16610 LV
Launch year: **2003**
Production quantity: **128,835**
Material: **STAINLESS STEEL**
Diameter: **40 MM**
Calibre: **3135**
Function: **HOURS, MINUTES, SECONDS AND DATE**
Waterproofness: **300 M/1,000 FT**

SUBMARINER REF. 116618 LN
Launch year: **2008**
Production quantity: **7,113**
Material: **18 CT YELLOW GOLD**
Diameter: **40 MM**
Calibre: **3135**
Function: **HOURS, MINUTES, SECONDS AND DATE**
Waterproofness: **300 M/1,000 FT**

SUBMARINER REF. 116618 LB
Launch year: **2008**
Production quantity: **11,666**
Material: **18 CT YELLOW GOLD**
Diameter: **40 MM**
Calibre: **3135**
Function: **HOURS, MINUTES, SECONDS AND DATE**
Waterproofness: **300 M/1,000 FT**

SUBMARINER REF. 116619 LB
Launch year: **2008**
Production quantity: **13,630**
Material: **18 CT WHITE GOLD**
Diameter: **40 MM**
Calibre: **3135**
Function: **HOURS, MINUTES, SECONDS AND DATE**
Waterproofness: **300 M/1,000 FT**

ROLEX DEEPSEA REF. 116660
Launch year: **2008**
Production quantity: **123,588**
Material: **STAINLESS STEEL**
Diameter: **44 MM**
Calibre: **3135**
Function: **HOURS, MINUTES, SECONDS AND DATE**
Waterproofness: **3,900 M/12,800 FT**

SUBMARINER REF. 116613 LN
Launch year: **2009**
Production quantity: **104,932**
Material: **YELLOW ROLESOR**
Diameter: **40 MM**
Caliber: **3135**
Function: **HOURS, MINUTES, SECONDS AND DATE**
Waterproofness: **300 M/1,000 FT**

SUBMARINER REF. 116613 LB
Launch year: **2009**
Production quantity: **143,264**
Material: **YELLOW ROLESOR**
Diameter: **40 MM**
Calibre: **3135**
Function: **HOURS, MINUTES, SECONDS AND DATE**
Waterproofness: **300 M/1,000 FT**

SUBMARINER REF. 116610 LN
Launch year: **2010**
Production quantity: **414,905**
Material: **STAINLESS STEEL**
Diameter: **40 MM**
Caliber: **3135**
Function: **HOURS, MINUTES, SECONDS AND DATE**
Waterproofness: **300 M/1,000 FT**

SUBMARINER REF. 116610 LV
Launch year: **2010**
Production quantity: **228,710**
Material: **STAINLESS STEEL**
Diameter: **40 MM**
Caliber: **3135**
Function: **HOURS, MINUTES, SECONDS AND DATE**
Waterproofness: **300 M/1,000 FT**

SUBMARINER REF. 116649 EMBR
Launch year: **2010**
Production quantity: **51**
Material: **18 CT WHITE GOLD**
Diameter: **40 MM**
Calibre: **3135**
Function: **HOURS, MINUTES, SECONDS AND DATE**
Waterproofness: **300 M/1,000 FT**

SUBMARINER REF. 114060
Launch year: **2012**
Production quantity: **153,437**
Material: **STAINLESS STEEL**
Diameter: **40 MM**
Calibre: **3130**
Function: **HOURS, MINUTES, SECONDS**
Waterproofness: **300 M/1,000 FT**

SEA-DWELLER REF. 116600
Launch year: **2014**
Production quantity: **27,102**
Material: **STAINLESS STEEL**
Diameter: **40 MM**
Calibre: **3135**
Function: **HOURS, MINUTES, SECONDS AND DATE**
Waterproofness: **1,220 M/4,000 FT**

SEA-DWELLER REF. 126600
Launch year: **2017**
Production quantity: **115,308**
Material: **STAINLESS STEEL**
Diameter: **43 MM**
Calibre: **3235**
Function: **HOURS, MINUTES, SECONDS AND DATE**
Waterproofness: **1,220 M/4,000 FT**

SUBMARINER REF. 116659 SABR
Launch year: **2018**
Production quantity: **771**
Material: **18 CT WHITE GOLD**
Diameter: **40 MM**
Calibre: **3135**
Function: **HOURS, MINUTES, SECONDS AND DATE**
Waterproofness: **300 M/1,000 FT**

ROLEX DEEPSEA REF. 126660
Launch year: **2018**
Production quantity: **88,571**
Material: **STAINLESS STEEL**
Diameter: **44 MM**
Calibre: **3235**
Function: **HOURS, MINUTES, SECONDS AND DATE**
Waterproofness: **3,900 M/12,800 FT**

SEA-DWELLER REF. 126603
Launch year: **2019**
Production quantity: **46,215**
Material: **YELLOW ROLESOR**
Diameter: **43 MM**
Calibre: **3235**
Function: **HOURS, MINUTES, SECONDS AND DATE**
Waterproofness: **1,220 M/4,000 FT**

SUBMARINER REF. 124060
Launch year: **2020**
Production quantity: **82,382**
Material: **STAINLESS STEEL**
Diameter: **41 MM**
Calibre: **3230**
Function: **HOURS, MINUTES, SECONDS**
Waterproofness: **300 M/1,000 FT**

SUBMARINER REF. 126610 LN
Launch year: **2020**
Production quantity: **177,328**
Material: **STAINLESS STEEL**
Diameter: **41 MM**
Calibre: **3235**
Function: **HOURS, MINUTES, SECONDS AND DATE**
Waterproofness: **300 M/1,000 FT**

SUBMARINER REF. 126610 LV
Launch year: **2020**
Production quantity: **103,584**
Material: **STAINLESS STEEL**
Diameter: **41 MM**
Calibre: **3235**
Function: **HOURS, MINUTES, SECONDS AND DATE**
Waterproofness: **300 M/1,000 FT**

SUBMARINER REF. 126613 LN
Launch year: **2020**
Production quantity: **54,092**
Material: **YELLOW ROLESOR**
Diameter: **41 MM**
Calibre: **3235**
Function: **HOURS, MINUTES, SECONDS AND DATE**
Waterproofness: **300 M/1,000 FT**

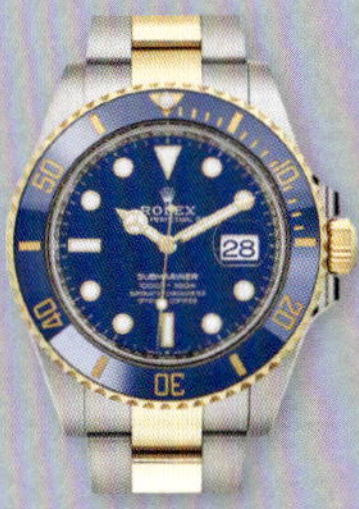

SUBMARINER REF. 126613 LB
Launch year: **2020**
Production quantity: **78,540**
Material: **YELLOW ROLESOR**
Diameter: **41 MM**
Calibre: **3235**
Function: **HOURS, MINUTES, SECONDS AND DATE**
Waterproofness: **300 M/1,000 FT**

SUBMARINER REF. 126618 LN
Launch year: **2020**
Production quantity: **6,663**
Material: **18 CT YELLOW GOLD**
Diameter: **41 MM**
Calibre: **3235**
Function: **HOURS, MINUTES, SECONDS AND DATE**
Waterproofness: **300 M/1,000 FT**

SUBMARINER REF. 126618 LB
Launch year: **2020**
Production quantity: **11,219**
Material: **18 CT YELLOW GOLD**
Diameter: **41 MM**
Calibre: **3235**
Function: **HOURS, MINUTES, SECONDS AND DATE**
Waterproofness: **300 M/1,000 FT**

SUBMARINER REF. 126619 LB
Launch year: **2020**
Production quantity: **11,464**
Material: **18 CT WHITE GOLD**
Diameter: **41 MM**
Calibre: **3235**
Function: **HOURS, MINUTES, SECONDS AND DATE**
Waterproofness: **300 M/1,000 FT**

ROLEX DEEPSEA REF. 136660
Launch year: **2022**
Production quantity: **27,025**
Material: **STAINLESS STEEL**
Diameter: **44 MM**
Calibre: **3235**
Function: **HOURS, MINUTES, SECONDS AND DATE**
Waterproofness: **3,900 M/12,800 FT**

DEEPSEA CHALLENGE REF. 126067
Launch year: **2022**
Production quantity: **1,298**
Material: **TITANIUM**
Diameter: **50 MM**
Calibre: **3230**
Function: **HOURS, MINUTES, SECONDS**
Waterproofness: **11,000 M/36,090 FT**

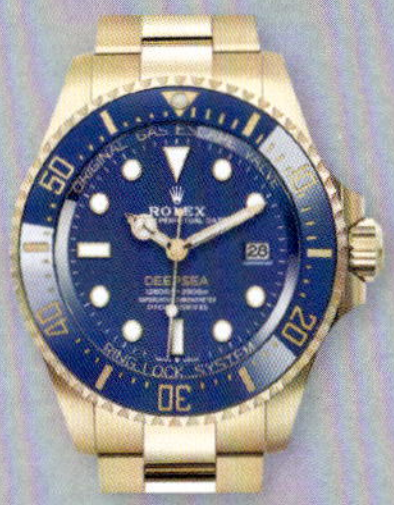

ROLEX DEEPSEA REF. 136668 LB
Launch year: **2024**
Production quantity: **UNDETERMINED AT TIME OF PUBLICATION**
Material: **18 CT YELLOW GOLD**
Diameter: **44 MM**
Calibre: **3235**
Function: **HOURS, MINUTES, SECONDS AND DATE**
Waterproofness: **3,900 M/12,800 FT**

The watches illustrated in the photographs are representative versions of each reference. Production quantities are estimates based on figures from the Rolex Archives and include all combinations of a given reference. Quantities related to references that remain in the catalogue will change over the years.

A SELECTION OF ADVERTISEMENTS FOR THE SUBMARINER

1954

1967

1972

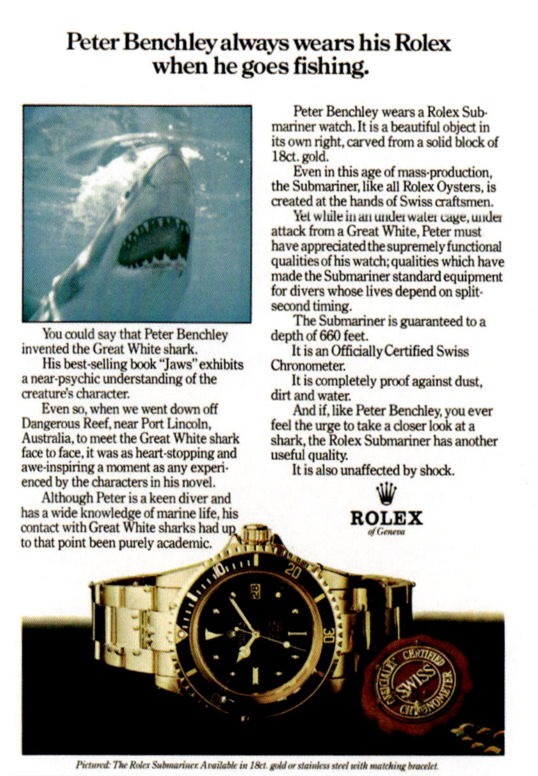

1976

1985

1986

1990

1995

1997

1999

2001

2002

2004

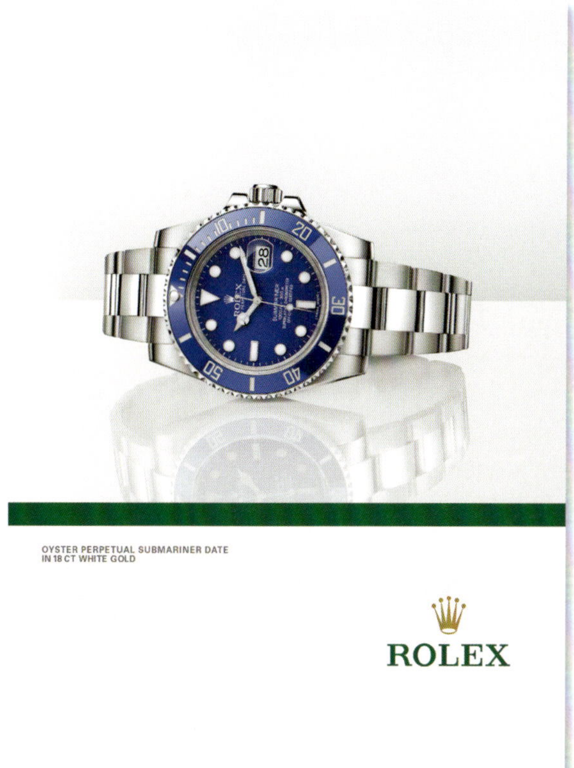

2012

2018

2020

THE NEW GENERATION #SUBMARINER #PERPETUAL

Submariner, 2020, Ref. 124060, 41 mm, Oystersteel,
waterproof to 300 metres (1,000 feet).

ROLEX
SUBMARINER

ENDNOTES

INTRODUCTION: THE VISION OF HANS WILSDORF

1. www.rolex.org
2. Speech given by Hans Wilsdorf at his 75th birthday dinner, Rolex Archives

CHAPTER I: A NEW CHAPTER IN WATCHMAKING HISTORY

1. Hugh Walker's letter from J. Walter Thompson – René-Paul Jeanneret, 29 January 1952, Rolex Archives
2. René-Paul Jeanneret, biographical note, 30 March 1973, Rolex Archives
3. Rolex to the Commanding Officer of the Royal Navy Diving School, 6 February 1952, Rolex Archives
4. Report by Mr Sheldrick on his visit to HMS *Reclaim*, 14 March 1952, Rolex Archives
5. Report by Mr Sheldrick on his visit to HMS *Reclaim*, 14 March 1952, Rolex Archives
6. Lieutenant Commander Bathurst Royal Navy – Rolex, 21 May 1952, Rolex Archives
7. Lieutenant Commander Bathurst Royal Navy – Rolex, 21 May 1952, Rolex Archives
8. René-Paul Jeanneret – Richard A. Winter, 24 May 1952, Rolex Archives
9. Richard A. Winter – René-Paul Jeanneret, 30 May 1952, Rolex Archives
10. Richard A. Winter – René-Paul Jeanneret, 30 May 1952, Rolex Archives
11. Richard A. Winter – The Secretary of the Admiralty, 30 May 1952, Rolex Archives
12. Richard A. Winter – René-Paul Jeanneret, 30 May 1952, Rolex Archives
13. Richard A. Winter – Rolex Watches, 18 September 1952, Rolex Archives
14. Richard A. Winter – Rolex Watches, 2 October 1952, Rolex Archives
15. Richard A. Winter – Lt Cdr Gutteridge, 15 October 1952, Rolex Archives
16. Richard A. Winter – Lt Cdr Gutteridge, 27 October 1952, Rolex Archives
17. Richard A. Winter – Lt Cdr Gutteridge, 6 October 1954, Rolex Archives
18. Correspondence between I. D. Johnstone and Lt Cdr T. McAndrew, MoD, 22 May 1972, Rolex Archives
19. Richard A. Winter – Lt Cdr Gutteridge, 6 October 1954, Rolex Archives
20. Richard A. Winter – Lt Cdr Gutteridge, 6 October 1954, Rolex Archives

CHAPTER II: A DIVER'S DREAM COME TRUE

1. Hans Wilsdorf – René-Paul Jeanneret, San Remo, 12 January 1953, Rolex Archives
2. Hans Wilsdorf – René-Paul Jeanneret, San Remo, 12 January 1953, Rolex Archives
3. Dimitri Rebikoff: *Free Diving*. Sidgwick & Jackson, London, 1955, p. 59
4. *National Geographic Magazine*, October 1952
5. Dimitri Rebikoff: *Free Diving*. Sidgwick & Jackson, London, 1955, p. 100
6. Dimitri Rebikoff: *Free Diving*. Sidgwick & Jackson, London, 1955, p. 100
7. *Historical Diver*, No. 21 Fall 1999, p. 35
8. Dimitri Rebikoff: *Free Diving*. Sidgwick & Jackson, London, 1955, p. 100
9. Dimitri Rebikoff: *Free Diving*. Sidgwick & Jackson, London, 1955, p. 177
10. *Popular Mechanics*, December 1967
11. René-Paul Jeanneret – Dimitri Rebikoff, Geneva, 3 March 1953, Rolex Archives
12. René-Paul Jeanneret – Xavier Fructus, 13 September 1954, Rolex Archives
13. Dimitri Rebikoff: *Free Diving*. Sidgwick & Jackson, London, 1955, p. 80
14. Dimitri Rebikoff – René-Paul Jeanneret, 7 April 1953, Rolex Archives
15. Dimitri Rebikoff – René-Paul Jeanneret, 7 April 1953, Rolex Archives
16. René-Paul Jeanneret – Dimitri Rebikoff, 13 April 1953, Rolex Archives
17. Dimitri Rebikoff – René-Paul Jeanneret, 1 June 1953, Rolex Archives
18. René-Paul Jeanneret – Dimitri Rebikoff, 28 May 1953, Rolex Archives
19. Dimitri Rebikoff – René-Paul Jeanneret, 1 June 1953, Rolex Archives
20. René-Paul Jeanneret – Dimitri Rebikoff, 28 May 1953, Rolex Archives
21. René-Paul Jeanneret – Dimitri Rebikoff, 28 May 1953, Rolex Archives
22. René-Paul Jeanneret – Dimitri Rebikoff, 28 May 1953, Rolex Archives
23. Dimitri Rebikoff – René-Paul Jeanneret, 1 June 1953, Rolex Archives
24. Technical conference, 28 May 1953, Rolex Archives
25. Report by Dimitri Rebikoff, 26 October 1953, Rolex Archives
26. Report by Dimitri Rebikoff, 26 October 1953, Rolex Archives
27. Report by Dimitri Rebikoff, 26 October 1953, Rolex Archives
28. Report by Dimitri Rebikoff, 26 October 1953, Rolex Archives
29. Report by Dimitri Rebikoff, 26 October 1953, Rolex Archives
30. Hans Wilsdorf – René-Paul Jeanneret, San Remo, 12 January 1953, Rolex Archives

CHAPTER III: AT THE CENTRE OF THE WORLD OF DIVING

1. Dr Alberto Grazioli – René-Paul Jeanneret, 31 October 1952, Rolex Archives
2. René-Paul Jeanneret – Marie Meinen, 12 November 1952, Rolex Archives
3. Memo to Ms Foex, 17 March 1954, Rolex Archives
4. Dr Luigi Stuart Tovini – Rolex Rome, 19 November 1953, Rolex Archives
5. Bertrand Gros, interview with the author, December 2020
6. *The Turn-O-Graph*, Rolex Archives
7. Pierre Cirier, Tunis, 15 November 1954, Rolex Archives
8. Pierre Cirier, Tunis, 15 November 1954, Rolex Archives
9. Henri Broussard – Rolex, Geneva, 4 February 1955, Rolex Archives
10. Henri Broussard – Rolex, Geneva, 4 February 1955, Rolex Archives

CHAPTER IV: EXPLORING THE SILENT WORLD

1. Dimitri Rebikoff: *Free Diving*. Sidgwick & Jackson, London, 1955, p. 9
2. Dimitri Rebikoff: *Free Diving*. Sidgwick & Jackson, London, 1955, p. 9
3. Dimitri Rebikoff: *Free Diving*. Sidgwick & Jackson, London, 1955, p. 9
4. Dimitri Rebikoff: *Free Diving*. Sidgwick & Jackson, London, 1955, p. 9
5. Dimitri Rebikoff: *Free Diving*. Sidgwick & Jackson, London, 1955, pp. 9–10
6. Dimitri Rebikoff: *Free Diving*. Sidgwick & Jackson, London, 1955, p. 9
7. René-Paul Jeanneret – Dimitri Rebikoff, 24 August 1953, Rolex Archives
8. René-Paul Jeanneret – Xavier Fructus, 13 September 1954, Rolex Archives
9. Xavier Fructus – René-Paul Jeanneret, 17 September 1954, Rolex Archives

Perpetual Planet Initiative: Rolex Testimonee Sylvia Earle

1. *Time*, 5 October 1998
2. *People Magazine*, 3 June 2000
3. *The New York Times Magazine*, 26 June 1991
4. *National Geographic Traveler*, February 1999
5. Sylvia Earle, email interview with the author, 2019
6. *The New York Times Magazine*, 26 June 1991
7. Sylvia Earle, email interview with the author, 2019
8. Sylvia Earle, email interview with the author, 2019
9. Sylvia Earle, email interview with the author, 2019

CHAPTER V: A NEW ERA OF DEEP-SEA EXPLORATION

1. Jacques Piccard and Robert S. Dietz: *Seven Miles Down: The Story of the Bathyscaph Trieste*. Longmans, London, 1962, p. 31
2. Jacques Piccard and Robert S. Dietz: *Seven Miles Down: The Story of the Bathyscaph Trieste*. Longmans, London, 1962, p. 34
3. Prof Auguste Piccard, Geneva, 25 February 1952, Rolex Archives
4. René-Paul Jeanneret – Prof Auguste Piccard, 29 February 1952, Rolex Archives
5. Prof Auguste Piccard – Rolex, Trieste, 1 April 1952, Rolex Archives
6. René-Paul Jeanneret – Prof Auguste Piccard, 23 April 1952, Rolex Archives
7. Prof Auguste Piccard – Rolex, 1 October 1953, Rolex Archives
8. *Revue Francaise des Bijoutiers Horlogers*, July 1954, Rolex Archives
9. Rolex – Mr Favre, Musée d'Horlogerie, 10 April 1957, Rolex Archives
10. Letter from Mr Favre, Musée d'Horlogerie – Rolex, 13 April 1957, Rolex Archives
11. United States Information Service Press Release, Bern, 13 September 1960, Rolex Archives
12. Jacques Piccard and Robert S. Dietz: *Seven Miles Down: The Story of the Bathyscaph Trieste*. Longmans, London, 1962, p. 17
13. Jacques Piccard and Robert S. Dietz: *Seven Miles Down: The Story of the Bathyscaph Trieste*. Longmans, London, 1962, p. 17
14. *Life*, 15 February 1960
15. *Life*, 15 February 1960
16. *Life*, 15 February 1960
17. *Life*, 15 February 1960
18. Jacques Piccard and Robert S. Dietz: *Seven Miles Down: The Story of the Bathyscaph Trieste*. Longmans, London, 1962, p. 162
19. *Life*, 15 February 1960
20. Jacques Piccard – Rolex, 21 January 1960, Rolex Archives

CHAPTER VI: DAY OF THE AQUANAUTS

1. Jacques Piccard and Robert S. Dietz: *Seven Miles Down: The Story of the Bathyscaph Trieste*. Longmans, London, 1962, p. 172
2. Jacques Piccard and Robert S. Dietz: *Seven Miles Down: The Story of the Bathyscaph Trieste*. Longmans, London, 1962, p. 183
3. Ben Hellwarth, *Sealab: America's Forgotten Quest to Live and Work on the Ocean Floor*, Simon & Schuster, New York, 2012
4. Joseph A. Tainter and Tadeusz W. Patzek: *Drilling Down: The Gulf Oil Debacle and Our Energy Dilemma*. Springer, New York, 2012, p. 222
5. Sherri Ferguson: "Saturation Diving, a quick and dirty guide", *Alert Diver*, online, 2017
6. Sherri Ferguson: "Saturation Diving, a quick and dirty guide", *Alert Diver*, online, 2017
7. Helen A. Siiteri: *Papa Topside: The Sealab Chronicles of Capt George F. Bond USN*. Naval Institute Press, Annapolis Maryland, 1993
8. Helen A. Siiteri: *Papa Topside: The Sealab Chronicles of Capt George F. Bond USN*. Naval Institute Press, Annapolis Maryland, 1993
9. *The New York Times*, 6 January 1983

10. Helen A. Siiteri: *Papa Topside: The Sealab Chronicles of Capt George F. Bond USN.* Naval Institute Press, Annapolis Maryland, 1993
11. Presentation, Invention No. 492246, Brevet Principal, 6 November 1967, Rolex Archives
12. Presentation, Invention No. 492246, Brevet Principal, 6 November 1967, Rolex Archives
13. Report on the conference held at the end of September 1966, Concerning the New Models, Rolex Archives
14. Report on the conference held at the end of September 1966, Concerning the New Models, Rolex Archives
15. Memo from the Styling Commission – Mr Perret, 7 June 1967, Rolex Archives

CHAPTER VII: THE NEXT FRONTIER

1. René-Paul Jeanneret – Col J. H. Glenn, 27 October 1967, Rolex Archives
2. Scott Carpenter – René-Paul Jeanneret, 29 November 1967, Rolex Archives
3. *Life,* 15 October 1965
4. *Life,* 15 October 1965
5. Letter from Scott Carpenter to René Girod, Rolex, New York, 27 February 1968, Rolex Archives
6. Jackson Tomsky, Department of the Navy Deep Submergence Systems Project Office – J. C. Miller (Rolex New York), 21 April 1969, Rolex Archives
7. www.rolexmagazine.com/2019/02/pbs-american-experience-sealab.html

Perpetual Planet Initiative: Rolex Testimonee Joe MacInnis

1. Victor R. Royce – John Nimmo (Rolex Geneva), 24 July 1967, Rolex Archives
2. René-Paul Jeanneret – Victor R. Royce (Rolex Toronto), 8 August 1967, Rolex Archives
3. Joe MacInnis: *Deep Leadership: Essential Insights from High-Risk Environments.* Alfred A. Knopf, Canada, 2012, p. 15
4. Joe MacInnis: *Deep Leadership: Essential Insights from High-Risk Environments.* Alfred A. Knopf, Canada, 2012, p. 15
5. *Popular Mechanics,* February 1969
6. Joe MacInnis – J. C. Miller (Rolex New York), 13 May 1968, Rolex Archives
7. Joe MacInnis – J. C. Miller (Rolex New York), 13 May 1968, Rolex Archives
8. Victor R. Royce (Rolex Canada) – John Nimmo (Rolex New York), 10 April 1969, Rolex Archives
9. Victor R. Royce (Rolex Canada) – John Nimmo (Rolex New York), 10 April 1969, Rolex Archives
10. Victor R. Royce (Rolex Canada) – John Nimmo (Rolex New York), 10 April 1969, Rolex Archives
11. Joe MacInnis – Victor R. Royce (Rolex Canada), 6 March 1970, Rolex Archives
12. Joe MacInnis – Victor R. Royce (Rolex Canada), 6 March 1970, Rolex Archives
13. Joe MacInnis – Victor R. Royce (Rolex Canada), 6 March 1970, Rolex Archives
14. Joe MacInnis – Victor R. Royce (Rolex Canada), 6 March 1970, Rolex Archives
15. Victor R. Royce – René-Paul Jeanneret (Rolex Geneva), 18 March 1970, Rolex Archives.
16. Joe MacInnis – Victor R. Royce (Rolex Canada), 6 March 1970, Rolex Archives
17. Joe MacInnis: *Underwater Images.* McClelland and Stewart, Toronto, 1971
18. *The First MacInnis Underwater Arctic Expedition August 1970,* Report, Rolex Archives

CHAPTER VIII: UNDER PRESSURE

1. *Rolex World,* No. 5 September 1975, Rolex Archives
2. *Rolex World,* No. 5 September 1975, Rolex Archives
3. *Rolex World,* No. 11 February 1978, p. 5, Rolex Archives
4. Rolex advertisement, 1972, Rolex Archives
5. Rolex advertisement, 1972, Rolex Archives
6. Rolex advertisement, circa 1982, Rolex Archives
7. Rolex advertisement, 1980, Rolex Archives
8. Rolex advertisement, circa 1987, Rolex Archives

CHAPTER IX: FROM HYCO TO THE HYDROGRAPHIC DEPARTMENT

1. Contemporary notes, 1970s, Rolex Archives
2. *Maclean's,* 1 March 1969
3. Contemporary notes, 1970s, Rolex Archives
4. René-Paul Jeanneret – Roland Puton, 17 November 1967, Rolex Archives
5. René-Paul Jeanneret – Donald Sorte, 16 February 1968, Rolex Archives
6. René-Paul Jeanneret – Donald Sorte, 16 February 1968, Rolex Archives
7. D. J. McDonald – Roland Puton, 23 April 1968, Rolex Archives
8. René-Paul Jeanneret – Victor R. Royce, 28 April 1971, Rolex Archives
9. Victor R. Royce – René-Paul Jeanneret, 27 July 1971, Rolex Archives
10. Victor R. Royce – René-Paul Jeanneret, 27 July 1971, Rolex Archives
11. Victor R. Royce – René-Paul Jeanneret, 27 July 1971, Rolex Archives
12. René-Paul Jeanneret – Victor R. Royce, 16 August 1971, Rolex Archives
13. D. J. McDonald – Victor R. Royce, 9 November 1971, Rolex Archives
14. Internal memorandum, 8 August 1973, Rolex Archives
15. Reproduced in *Rolex World,* No. 18, March 1981, Rolex Archives
16. I. D. Johnstone – Lt Cdr T. McAndrew, MoD, 22 May 1972, Rolex Archives
17. W. P. Roseman – I. D. Johnstone, 29 March 1972, Rolex Archives
18. René-Paul Jeanneret – I. D. Johnstone, 10 April 1972, Rolex Archives
19. René-Paul Jeanneret – I. D. Johnstone, 10 April 1972, Rolex Archives
20. I. D. Johnstone – Hermann Van Gessel, 4 May 1972, Rolex Archives
21. I. D. Johnstone – Hermann Van Gessel, 4 May 1972, Rolex Archives
22. I. D. Johnstone – Hermann Van Gessel, 4 May 1972, Rolex Archives

CHAPTER X: SECRET AGENT, FAMOUS WATCH

1. Rolex advertisement, circa 1966, Rolex Archives
2. Foire Oceanographique Munich 1969, Rolex Archives
3. Foire Oceanographique Munich 1969, Rolex Archives
4. Ian Fleming: *Thunderball.* Pan Books, 1963, p. 125
5. Ian Fleming: *On Her Majesty's Secret Service.* Pan Books, 1964, p. 181

CHAPTER XI: THE SUBMARINER ON SCREEN

1. Rolex – Egil Woxholt, Remus Films Ltd., 11 June 1957, Rolex Archives
2. Richard A. Winter – René-Paul Jeanneret, 14 June 1957, Rolex Archives
3. Richard A. Winter – René-Paul Jeanneret, 14 June 1957, Rolex Archives
4. Richard A. Winter – René-Paul Jeanneret, 14 June 1957, Rolex Archives
5. Vic Betts, Remus Films Ltd. – Rolex, 3 July 1957, Rolex Archives
6. Laurence Harvey – René-Paul Jeanneret, 30 September 1957, Rolex Archives
7. *The New York Times,* 21 June 1975
8. Carl Gottlieb: *The Jaws Log: 30th Anniversary Edition.* Faber & Faber, London, 2005
9. Rolex advertisement featuring Peter Benchley, circa 1975, Rolex Archives
10. Rolex advertisement featuring Peter Benchley, circa 1976, Rolex Archives
11. Rolex advertisement, circa 1976, Rolex Archives
12. *The New York Times,* 3 July 1977
13. *The New York Times,* 3 July 1977
14. *The New York Times,* 3 July 1977
15. *The New York Times,* 3 July 1977
16. *The New York Times,* 3 July 1977
17. Rolex advertisement, end of the 1970s, Rolex Archives
18. J. Walter Thompson's Rolex International Bulletin, No. 7, August 1977, Rolex Archives
19. J. Walter Thompson's Rolex International Bulletin, No. 7, August 1977, Rolex Archives
20. Klarlund – Rolex, 5 October 1977, Rolex Archives

CHAPTER XII: AN IMMUTABLE ICON

Perpetual Planet Initiative: Rolex Testimonee David Doubilet

1. Interview in *Perpetual Spirit*
2. Interview in *Perpetual Spirit*
3. www.rolex.org/environment/master-photographer-on-a-mission-to-help-us-see-the-sea

CHAPTER XIII: THE PAST TWENTY YEARS

1. C. Calvani, interview with the author, December 2017
2. Hans Wilsdorf – René-Paul Jeanneret, San Remo, 12 January 1953, Rolex Archives
3. *The Oyster: 90 Years of Perpetual Innovation,* Baselworld 2016, Rolex Archives
4. www.rolex.com/about-rolex-watches/a-watch-like-me.html
5. www.rolex.com/about-rolex-watches/a-watch-like-me.html
6. *Rolex Deepsea Challenge, to the deepest reaches of the oceans,* Rolex Archives
7. *Rolex Deepsea Challenge, to the deepest reaches of the oceans,* Rolex Archives
8. *Rolex Deepsea Challenge, to the deepest reaches of the oceans,* Rolex Archives
9. *Rolex Deepsea Challenge, to the deepest reaches of the oceans,* Rolex Archives
10. *Rolex Deepsea Challenge, to the deepest reaches of the oceans,* Rolex Archives
11. *Rolex Deepsea Challenge, to the deepest reaches of the oceans,* Rolex Archives
12. *National Geographic,* 25 March 2012
13. *Rolex Deepsea Challenge, to the deepest reaches of the oceans,* Rolex Archives

Perpetual Planet Initiative: A Family Of Ocean Heroes

1. www.rolex.org/environment/perpetual-planet/cristina-mittermeier
2. *The Rolex Magazine,* Issue 11
3. *The Rolex Magazine,* Issue 11
4. *The Rolex Magazine,* Issue 11
5. *The Rolex Magazine,* Issue 11
6. *The Rolex Magazine,* Issue 11
7. www.paulnicklen.com/fineart/killer-tension/
8. www.rolex.org/environment/perpetual-planet/cristina-mittermeier
9. https://www.seattleaquarium.org/stories/qa-cristina-mittermeier/
10. Email to the author, September 2023

EPILOGUE

1. *National Geographic,* 25 March 2012
2. *National Geographic,* 25 March 2012

CREDITS

Front cover: ©Rolex/JVA Studios
Back cover: ©Rolex/Kip Evans
All images ©Rolex with the exception of: pp. 26, 29, 35, 48, 72–73, 74, 90–91, 100, 111, 112–113, 119, 125, 137, 140–141, 163, 169, 170–171, 188–189, 190, 192–193, 198–199, 227, 229, 232–233, 235, 246–247 ©Wallpaper*/Kate Jackling; pp. 4–5 ©Aaron Foster/Getty Images; pp. 6–7 ©Rolex/Stefan Walter; pp. 8–9, 147 ©Rolex/JVA Studios; pp. 10–11, 60–61 (right), 212 (bottom left, bottom right), 213 (bottom), 224–225, 226, 230–231 ©Rolex/Franck Gazzola; pp. 17, 44, 47, 69, 92–93, 168 ©Rolex/Jean-Daniel Meyer; p. 18 ©Central Press/Stringer/Getty Images; pp. 18–19 (middle) ASSOCIATED PRESS Photo/Charles Kelly/Alamy; p. 19 ©Pablo Picasso (b/w photo)/Private collection/; ©Estate of Eddy Novarro. All rights reserved 2023/; Bridgeman Images ©Succession Picasso/DACS, London; pp. 20, 52, 55 (bottom), 58, 60, 86, 98–99 (left), 128, 139, 144, 155 (left) ©Alamy; p. 25 (top and bottom) ©IWM; p. 30 © Rolex, courtesy of Dimitri Rebikoff; p. 33 ©Travers, courtesy of Dimitri Rebikoff; p. 34 ©Henri Broussard, courtesy of Dimitri Rebikoff; p. 50 ©Rolex/Régis Golay; pp. 55 (top), 71 (top), 81, 94 ©Bettmann/Getty Images; pp. 56–57 ©ABC Photo Archives/Disney General Entertainment Content via Getty Images; p. 59 ©Granger/Shutterstock; pp. 62–63 ©Al Giddings; pp. 64–65 ©Rolex/Kip Evans; p. 66 ©Thomas J. Abercrombie/*National Geographic*; p. 76 (left) ©Keystone/Len Sirman -Archive/Stringer; pp. 76–77 (right) ©Black Star, New York; p. 82 ©Comex; p. 85 ©Xavier DESMIER/Gamma-Rapho via Getty Images; p. 97 ©Jonathan William Mitchell/Age Foto stock; p. 99 (right) ©Rolls Press/Popperfoto via Getty Images; pp. 102–103 ©Frank Lennon/*Toronto Star* via Getty Images; p. 106 ©Rick Mason, courtesy of Joe Macinnis; p. 108 ©Rolex, courtesy of Comex; pp. 126, 130 ©Danjaq/Eon/Ua/Kobal/Shutterstock; p. 129 ©Bert Cann/Danjaq/Eon/Ua/ Kobal/Shutterstock; pp. 130–131 (right), 142–143 ©Popperfoto via Getty Images; p. 135 ©Bela Zola/*Daily Mirror*/Mirrorpix via Getty Images; p. 136 ©Snap/Shutterstock; p. 138 ©Rolex/Syd Cain; p. 142 (left) ©Eon Prods/Kobal/Shutterstock; p. 150–151 ©HA/THA/Shutterstock; p. 154 ©Mgm/Kobal/Shutterstock; p. 155 (right) ©Gene Page/Marvel/Columbia Tristar/Kobal/Shutterstock; p. 156 (left) ©Mallick Media/Kobal/Shutterstock; p. 156 (right) ©New Regency Pictures/Kobal/Shutterstock; p. 157 (left) ©Twentieth Century Fox Film Corporation/Kobal/Shutterstock; p. 157 (right) ©Moviestore/Shutterstock; p. 158 ©Shutterstock; pp. 160–161 ©Bernard Cahier/Getty Images; p. 167 ©Rolex/Thomas Hensinger; pp. 172–173 (left), 173 (right), 174–175 (middle) ©Rolex/ François Lacour; p. 174 (left) ©Rolex/Cédric Widmer; pp. 175 (right), 176–177, 210, 211 ©Rolex/Fred Merz; pp. 178–179 ©Rolex/ Juriaan Booij; p. 180 (left) ©Rolex/Bart Michiels; pp. 180–181 (right), 184–185 ©David Doubilet; pp. 182–183 ©Jennifer Hayes, courtesy of David Doubilet; pp. 186, 207, 208, 209 ©Mark Thiessen/*National Geographic*; p. 195 ©Ghislain Bardout/Deepsea Under The Pole by Rolex; pp. 196–197 (left) ©Benoît Poyelle/Deepsea Under The Pole by Rolex; p. 197 (right) ©Rolex/Francis Azémard; pp. 202 (left), 203 (right) ©Rolex/Jess Hoffman; pp. 202–203 (middle), 204 ©Rolex/Stojan&Voumard; p. 205 ©Rolex/Alain Costa; p. 212 (top left) ©Anna Heupel; p. 212 (top right), 223 ©Rolex/Tim McKenna; pp. 213 (top), 215, 216–217 ©Cristina Mittermeier; pp. 218–219 ©Jed Weingarten; pp. 220–221 ©Paul Nicklen

ACKNOWLEDGEMENTS

This is the first in a series of books, each of which is devoted to exploring and understanding one of the celebrated models of Rolex. The series starts with the Submariner, but it could equally have begun with the humble Oyster Perpetual or the celebrated Daytona. Each model has stories to tell. Much more than famous watches, they are among the most familiar man-made objects of our times. But just because we are familiar with them does not mean that we understand them and how they came to be what they are today. Even devoted collectors, their understanding mediated by serial numbers, crown diameters, case dimensions, minute cosmetic differences on the dial and the like, have had to assemble their information piecemeal from diverse sources. Now, for the first time, Rolex has opened its archives enabling the stories of these models to be told in a way that places them in the context of their times and recounts the events, some momentous, other apparently trivial, that have contributed to their development. These books are not intended to be a complete and exhaustive survey of every declension of every model (that would require many more pages and volumes), nor do they have any pretensions to omniscience: there are gaps still to be filled and mysteries yet to be elucidated. Instead, they are intended to enrich and enhance the way we experience a remarkable brand that has witnessed truly historic events and which has itself participated in the making of history. Above all Rolex is about people, those who wear the watches, those who make and sell them, and of course, Hans Wilsdorf, who founded the brand, and it is to those people, past, present and future, that this series of books is dedicated.

Nicholas Foulkes

AUTHOR BIO

Nicholas Foulkes is the author of around 40 books, many on watches, and others on subjects as diverse as the Battle of Waterloo and gambling in the 1840s. He has written biographies of Count d'Orsay, Bernard Buffet and Princess Ira von Fürstenberg. He is well known for his writing on material culture, the arts, jewellery and timepieces. Over the course of a career spanning four decades, he has contributed articles to numerous British and international newspapers, magazines and websites. A graduate of Hertford College, Oxford, he lives in London with his wife; they have two adult sons.

ROLEX

ROLEX SA
3–7, RUE FRANÇOIS-DUSSAUD
1211 GENEVA 26
SWITZERLAND

rolex.com
rolex.org
ISBN: 978-1-80521-893-7
Printed in Italy by Grafiche Milani
10 9 8 7 6 5 4 3 2

Malcolm Young – Wallpaper* Managing Director
Bill Prince – Wallpaper* Bespoke Editorial Director
SJ Molony – Wallpaper* Bespoke Director
Vicki Morris – Wallpaper* Bespoke Watches & Jewellery Director
Carly Gray – Wallpaper* Bespoke Producer
Anya Hassett – Wallpaper* Bespoke Producer
Esther Kremer – PRINT Editorial Director
Camille Dubois – PRINT Creative Director